# EARTH STORY

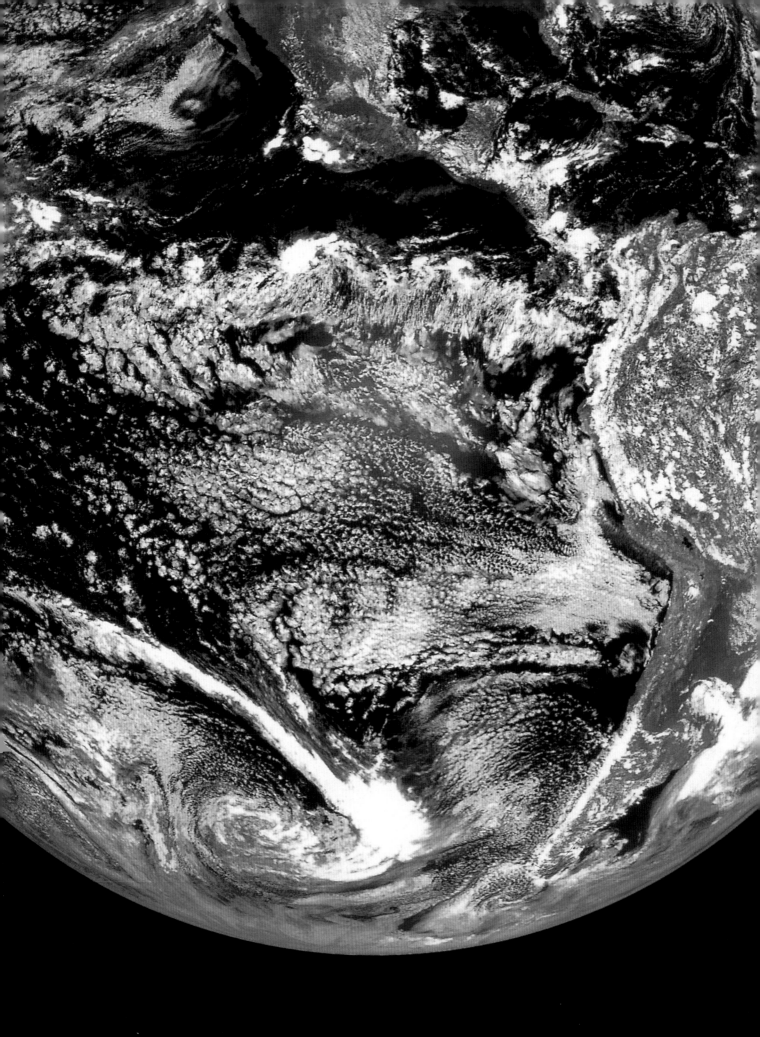

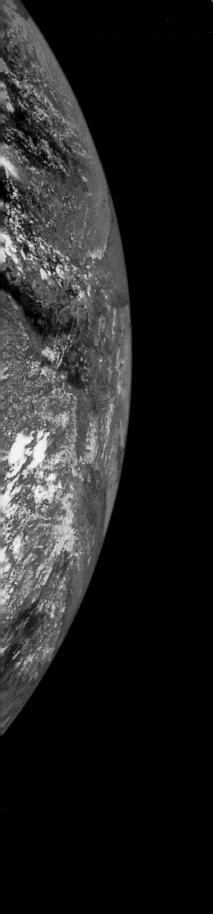

# earth STORY

## THE SHAPING OF OUR WORLD

SIMON LAMB
& DAVID SINGTON

BBC BOOKS

# Preface

*Earth Story*, television series and book, is the fruit of an unusual collaboration between a scientist with a long-standing interest in film-making (Simon Lamb) and a film-maker with a long-standing interest in the Earth Sciences (David Sington). The basic outline of *Earth Story* is the product of many conversations we have had over the years.

The task of writing the book, converting the series synopsis into a full-length manuscript, fell mainly to Simon. David and Simon worked together on this original draft, trying to ensure that it was accessible to as wide an audience as possible. Here the illustrations are obviously vital. Simon, working closely with Felicity Maxwell, created the scientific briefs for our superb illustrator, Gary Hincks (who was helped by Michael Eaton). Simon was also responsible for captioning and labelling the photographs and illustrations. The manuscript benefited from Felicity Maxwell's geological expertise, and she made numerous valuable suggestions. We would also like to thank Philip England, Rachel Mills, Keith O'Nions and Hazel Rossotti for comments on parts of the manuscript, and Hendrik van Heijst, Lorcan Kennan, Conall MacNiocall, Graham Robertson, Paul Valdes and John Woodhouse for creating the computer-generated diagrams.

This book is published to accompany the television series *Earth Story*
which was first broadcast on BBC 2 in 1998.
Executive Producer: Richard Reisz
Series Producer: David Sington
Producers: Robin Brightwell, Simon Lamb, Cynthia Page, Danielle Peck, Isabelle Rosin, Simon Singh

Published by BBC Books, an imprint of BBC Worldwide Ltd.,
Woodlands, 80 Wood Lane, London, W12 0TT

Reprinted 1998 (twice), 1999
© Simon Lamb and David Sington 1998
The moral right of the authors has been asserted

ISBN 0 563 38799 8

Illustrations by Gary Hincks
Additional illustrations by Michael Eaton
Illustration consultant: Felicity Maxwell
Designer: Ben Cracknell
Layout: Reuben Barnham

Commissioning editor: Sheila Ableman
Project editor: Martha Caute
Copy editor: Christine King
Art editor: Jane Coney
Picture researcher: Susannah Playfair

Set in Cheltenham Light
Printed and bound in Great Britain by Butler and Tanner, Frome & London
Colour separations by Radstock Reproductions Ltd, Midsomer Norton
Jacket printed by Lawrence Allen Ltd, Weston-super-Mare

*Frontispiece: A now familiar view of our planet from space, centred on the eastern Pacific Ocean. The deep blue oceans, white swirling clouds and green land areas would clearly show any alien space explorer that this planet is special, containing liquid water and life.*

*Page 6: Farewell to Earth. A picture of the Earth and Moon taken by the Galileo spacecraft in the 1990s as it embarked on its long journey to Jupiter.*

# Contents

· · · · · · · · · · · · · · · ·

# Introduction

In December 1990, as NASA's *Galileo* space probe left the vicinity of Earth on its way to Jupiter, it recorded a remarkable series of images of our planet with its attendant Moon. Against the black of outer space the vibrant blue and white patterns of our world contrast with the dun colour of its lifeless satellite. But *Galileo* did more than simply photograph the Earth. The spacecraft also performed a number of observations designed to answer an intriguing question of great significance for the future exploration of our galaxy: is it possible to detect from deep space the presence of living things on the surface of a planet? Does life send out any tell-tale signs across the Universe?

Imagining for the purposes of the experiment that the Earth was a newly discovered world about which they knew nothing, the NASA scientists were able to deduce from the light reaching their spacecraft the composition of the thin shell of gases that envelop the globe. Among the various gases they detected, they considered two of particular significance: water vapour and oxygen. Large areas of the planet appeared to be highly reflective, implying that the surface in these areas was liquid, and the presence of water in the atmosphere suggested that these oceans were made of water. Liquid water is an ideal medium for life: any planet that has it will be hospitable to living things. The presence of oxygen has equally profound implications. Oxygen is an extremely chemically reactive molecule, which means that it will readily combine with other elements. In this way all the oxygen will quickly disappear from a planet's atmosphere – unless, that is, there is some process able to replace it. Scientists know of only one likely candidate: the photosynthetic activity of living cells such as are found in plants. The NASA researchers concluded that the presence of both water and oxygen was a strong indicator for the existence of life.

The *Galileo* experiment should remind us of just what a remarkable place the Earth is. Of all the worlds that we know of, only our own reveals this twin signature of life. The Earth is unique in yet another respect, one which is made manifest not by the perspective of space but of time. Imagine for a moment a camera placed in orbit high above our world. Suppose too that this camera has been set up to record time-lapse photography of the globe below, with one frame taken every 50,000 years, say. The resulting film would show a planetary surface in constant motion. Continents would sail across the face of the Earth, colliding to form one great landmass, only to split up again into new and unfamiliar shapes. At the poles vast ice sheets would rapidly wax and wane, disappearing altogether for a while, and then suddenly reappearing. Mountain ranges would rise and fall. No other planet in the Solar System exhibits such frenetic geological activity.

In this book, and the accompanying television series, we contend that it is not merely coincidence that the planet which, uniquely, harbours life is also a place of ceaseless geological change. Indeed, we see the Earth's biological and geological activity as two sides of the same coin. *Earth Story* explores the emerging scientific realization that only a planet which is geologically active can sustain life, and only because the Earth is home to living things has it remained so geologically active. The argument is developed over the book's eight chapters, but it is worth summarizing here.

The opening chapter of our book describes how geologists learnt to read the rocks to reveal the history of the Earth. The story they have uncovered is one of great transformations, yet fundamental continuity. The oldest rocks on the planet, found in Greenland and southern Africa, suggest that some 4 billion years ago the world looked very different from today, covered by a single vast ocean dotted with thousands of volcanic islands. Yet the processes of volcanism, erosion and plate tectonics were already starting. Life too already had a foothold, in the form of heat-loving bacteria clustered around bubbling mud pools. The message from the rock record is that while the Earth has changed, the forces that shape it have not.

The nature of these forces is explored in the next three chapters. The key to understanding them lies in the basic laws of physics. The interior of the Earth is very hot, at a temperature of about 5000 Celsius. The surface of the planet, by contrast, is rather cool, and so heat will naturally flow out from the Earth's interior. The way that this heat moves is by being carried by the rock itself, in a process called convection. Hot rock from deep within the Earth rises towards the surface. Once there it moves horizontally sideways, gradually cooling before descending back towards the centre. The horizontal surface movement is described by the theory of plate tectonics, which explains how the outer shell of the Earth is divided into a small number of vast pieces, called plates, which are in constant motion relative to one another. Our planet therefore resembles a gigantic engine, converting the heat of the interior into the motion of the surface.

Over geological time plate movements have constantly shuffled the continents across the globe. In doing so, they have affected the planet's climate, as we describe in Chapters 5 and 6. The impact of climatic change on life was brought home to us when, during the filming of *Earth Story*, we flew to a scientific station in the middle of the great Greenland ice cap. The ice sheet itself is almost the negation of a landscape; an imperceptibly sloping dome 3000 metres high at its apex, it covers the hills and valleys below with a seemingly flat and featureless white plain, as if someone had erased the map of the world leaving nothing but a blank piece of paper. For life too, the ice cap is a virtual blank – there is no more sterile desert on the planet. The scientists who work at its centre are among the most isolated organisms on the globe. Nothing bigger than a microbe could make its permanent home in such a place. Yet, just a few thousand years ago, much of the northern hemisphere, a region that today supports a rich diversity of life, was covered with just such an ice cap.

Earth history is punctuated with many such shifts in climate, and as we discuss in our penultimate chapter, they are often associated with abrupt changes in the fossil record. It seems that events such as Ice Ages have had a profound influence on the course of evolution. The diversity of living things that we see today is a product of the Earth's restless dynamism.

Yet that dynamism may itself be sustained by life. Here we return to the *Galileo* probe and its 'discovery' of Earth's oceans. Water is clearly vital for life. What is perhaps more surprising is that water plays a crucial role in lubricating the motion of the plates – without it there would be no plate tectonics. So water quickens life and the Earth itself. But oceans can only exist within a very narrow temperature range, which must have been maintained throughout Earth history, despite the fact that ever since the Solar System formed 4.6 billion years ago, the amount of energy being put out by the Sun has been slowly increasing. Something prevented the Earth from overheating and losing its water by evaporation, the fate which seems to have befallen Venus.

Most scientists believe that the atmosphere of the early Earth was rich in carbon dioxide, exerting a powerful greenhouse effect to keep the infant planet warm. Today, most of that carbon dioxide is locked up in rocks in the crust, so that the planet is cooler now than it has been for most of its history, despite the Sun's greater output. The process of sequestering carbon in this way is mediated by living things, as they take carbon dioxide from the atmosphere to build their tissues. As more solar energy became available to life, it would have drawn down ever greater quantities of carbon from the atmosphere, thereby opening the windows of the greenhouse and negating the rise in temperature that would otherwise have occurred.

If this idea, which we discuss in our final chapter, is correct then we are forced to an astonishing conclusion. Without water there would be no life, but without life there would be no water, and without water there would be no plate tectonics; and so the Earth's geological and biological activity have been inextricably linked throughout its long history.

We human beings are the ultimate products of that history, truly children of the Earth. Under our gaze this extraordinary world, part machine, part organism, is beginning to yield up its secrets. It is our hope that as we understand more of our planet we may learn to live more wisely upon its surface, respecting the Earth as the thing that gave us birth.

David Sington

# THE TIME TRAVELLERS

•••••••••••••••••••••••••••••••••

Geologists, who study the Earth, seek to
understand the processes that have shaped our
planet throughout its history, creating the world we
see around us. To do so, they must reconstruct the
Earth's past. Yet how can we tell what happened in
distant epochs when there were no witnesses to
record events? Around 200 years ago scientists first
began to realize that clues to the past lay all around
them, in the rocks that make up the Earth's surface.
As they learnt how to read these rocks, they began
a journey back through time which geologists
continue to this day.

*Little Langdale Tarn, Cumbria, England: a timeless and unspoilt landscape. But geologists can
see the tell-tale signs of powerful forces – the scouring action of vast ice sheets, the erosion of rivers,
collapsing hillsides – which over geological time have profoundly altered this landscape.*

## LOOKING BENEATH THE LANDSCAPE

Imagine climbing to a high point on a clear day. Spread out before you is an intricate landscape of valleys and hills beneath a distant skyline of intersecting curves, outlining the tops of faraway mountains. Rivers, glinting in the sun, snake their way down the valley floors, meeting up with streams from other valleys to form a wider body of water which flows ever onwards to the sea. Everywhere is green vegetation: small clumps of trees cluster on the grassy river banks, and tall stands of conifers march up the sides of the hills. The air is full of bird song and the distant sounds of grazing sheep; in the foreground a fox darts across a field on its way to some mysterious rendezvous.

This imaginary vista probably accords with most people's idea of earthly paradise: a timeless landscape at ease with itself. But in reality, many features of this scene are probably very young. No doubt human activity has created the fields, felled the forests and husbanded some of the animals. In the distant past, many of the present animal and plant species would not have existed. More fundamentally, the rivers, which seem such an integral and peaceful part of the landscape, are powerful agents of change themselves. They have carved out the valleys and moulded the hills. Today, they are still wearing down the surface, carrying material away to build up a new landscape elsewhere. In other words, the surface of the Earth is not frozen in time, but contains a myriad of activities which are forever changing its shape. The landscape we see is like a single still frame in a film: if we could watch all the frames in succession, we would see a picture of continual motion.

This way of thinking about the world around us may be rather unfamiliar to most people, yet it is the basic viewpoint of any geologist. Essential to this perspective is an appreciation of the immense extent of Earth history. Armed with this knowledge, the geologist can see the true significance of small

changes such as scree falling down a steep hillside or a river shifting its banks during a flood. Given enough time, countless such tiny events have the power to transform the landscape into something quite different. This understanding of time started to emerge only 200 years ago when scientists began to realize that the history of the Earth could be reconstructed by a careful study of the rocks that lie at, or below, its surface.

It is still all too easy to take for granted the ground beneath our feet, without worrying too much about how it got there, treating it as solid 'stuff' which can be tunnelled into or built on. And yet one of the most extraordinary features of our planet is that this solid 'stuff' is so full of information, so rich in history, that to those who are interested it is virtually demanding to be heard, shouting out its story. Beneath us in the rocks is a record of almost everything that has happened on this planet, complete with dates, right back to a time when the Earth was a new feature of the Universe.

## READING THE ROCKS

Until the late seventeenth century, most European Christians believed the biblical creation story quite literally. The Book of Genesis outlined a timetable of events for Earth history, starting with the creation of the world in six days. In 1650, Archbishop James Ussher used the biblical chronology, adding up the lifespans of all the descendants of Adam, to calculate that the world was created in 4004 BC. This provided roughly 6000 years for the history of the Earth. Scientists who thought that Archbishop Ussher's estimate, or even ten times this, was an immense length of time were in for a rude shock when they started looking closely at the solid part of the Earth.

The late eighteenth century saw the beginning of the Industrial Revolution. Miners and engineers were acquiring an extensive knowledge of rocks, as they tunnelled through them to extract minerals or build canals. One such mining expert was Abraham Gottlob Werner, who worked in Freiburg in Saxony. He noticed that in this part of Germany there seemed to be a sequence in the rocks. The bottom of the sequence was made up of crystalline rocks, such as granite, gneiss and schist, found today in the Bohemian Massif. Resting on these was a succession of layered rocks, including sandstones and shale. Werner explained this by assuming that the Earth was originally completely enveloped by water – an idea strongly reminiscent of the great biblical Flood, recorded in the Book of Genesis. This primeval ocean varied in depth, with many deep and shallow parts. According to Werner, all the material that today makes up the outer crust of the Earth was originally either held in suspension or dissolved in the water, and through time, as the water subsided, this material precipitated or settled out to form the sequence of distinctive rock layers he had observed. First, the chemicals dissolved in the water crystallized out, coating the irregular bottom of the ocean with crystalline rocks. As time passed, the chemical precipitates became mixed with material which, originally in suspension, had now begun raining down on the bottom. Eventually, the material forming at the bottom of the ocean was predominantly suspended matter accumulating as successive layers. Finally, when the waters had subsided altogether, rivers flowing into lowland areas dumped on top of this sequence mud and sand derived from the disintegration of deposits which had formed in shallow parts of the original ocean.

In Werner's theory, almost all rocks are formed by what modern geologists think of as sedimentary processes, accumulating on the bottom of oceans, lakes or rivers. All those who subscribed to his theory were labelled 'Neptunist' because water played such a big role in their model. But the Neptunists were soon challenged by a rival theory, put forward by a Scottish farmer and businessman, James Hutton, who had made enough money to indulge a lifelong interest in the natural world. Hutton is justly celebrated for two observations he made in Scotland in the 1780s. The first was made in 1785 in Glen Tilt, where he found three different rock types: limestone, shale and granite. In Werner's theory, granite was a chemical precipitate from water. In Glen Tilt, however, the granite formed numerous fingers and veins which invaded the limestone and shale. Occasionally, blocks of shale were totally surrounded by granite. These observations convinced Hutton that the granite was not a chemical precipitate at all, but had once been molten rock which had intruded into older sedimentary rock – the limestone and shale.

Hutton's observations helped to convince him that heat as well as water had played an important role in the formation of the Earth's crust. He saw volcanoes, which he knew about from Italy and the Auvergne region of France, as 'a spiracle to the subterranean furnace' releasing the Earth's internal heat in the form of molten rock. The rock that makes up many of these volcanoes had been recognized as a lava called basalt. Hutton realized that some of this molten rock may not reach the surface, but may solidify inside the crust to form veins. Granite was merely another form of molten rock from the Earth's interior. Scientists who stressed the importance of volcanoes in the formation of the Earth's crust were labelled as 'Vulcanists'. Hutton's emphasis on molten rock at depth in the crust led to him being dubbed a 'Plutonist' after Pluto, the Greek god of the underworld. His recognition of the origin of granite was not a minor point, because so much surface rock, especially in the interiors of the major continents, is granite.

Like Hutton, geologists today recognize two principal types of rock: sedimentary rock, which usually forms layers made up of fragments of older rock that have been carried away by rivers and deposited in water; and rock which has cooled from a molten state. The latter is often referred to as igneous. Hutton also discovered a third important type of rock, though he makes no specific comment about it. This is metamorphic, rock which was originally either sedimentary or igneous, but has been subsequently altered by the effects of heat or pressure. For instance, the baked sedimentary rock that Hutton found at the margins of the once molten granite in Glen Tilt is a metamorphic rock.

But James Hutton was not just concerned with the origin of different rock types. In 1788, he made an observation which changed for ever the way geologists thought about the Earth, opening up undreamt-of vistas of time stretching into the remote past. At Siccar Point, on the east coast of Scotland near Edinburgh, he found a sequence of sandstones which was gently tilted. But the sandstones overlay a sequence of shales and siltstones in which the beds were nearly vertical and sometimes folded back on themselves. This relationship is often described as Hutton's unconformity, because it shows a dramatic break between two episodes of accumulation of sedimentary rock. Hutton worked out that only the following sequence of seven events could account for his observation:

1 Rivers eroded an ancient landscape, shifting fragments of the bedrock as sediment down to the sea.

2 The material carried by the rivers accumulated at the bottom of the sea to form a sequence of silts, shales and sands, which were buried and eventually became horizontal layers of rock.

3 These rock layers were uplifted out of the sea by movements inside the Earth. In the process they were turned from the horizontal to the vertical, contorted and folded back on themselves.

4 Rivers flowed off the uplifted and contorted rock, wearing down the surface to a flat plain.

5 Subsequently, the flat plain subsided and became the site of accumulation of a new sequence of sands, carried by rivers from high ground elsewhere.

6 Another period of Earth movements uplifted and tilted the new sequence of sediments.

7 Rivers today are again wearing away the uplifted rock, creating the present landscape on top.

This is a staggering amount of information to tease out of one outcrop of rock, and yet it is still a fraction of the information that can be extracted. A modern geologist would be able to work out which way the rivers were flowing, where in the sea the sediments were accumulating, the precise nature of the Earth movements that caused the uplift, and the age of the rocks. Even without this information, Hutton was able to deduce that there are many cycles in the formation of the Earth's surface, with an interplay between the degradation of the surface, as rivers wear away the landscape and carry material away, and growth, as accumulating sediment builds up the surface. Movements in the Earth's crust also play a vital role, periodically reversing the surface topography, so that lowlands become high and highlands become low.

Much of the power of Hutton's interpretation of the unconformity comes from the recognition that part of the bedrock is made up of sediments that were formed by processes, such as the action of rivers, that we can observe today. This simple but powerful notion cannot be found in any of the earlier theories of the Earth, or even Abraham Werner's Neptunist theory. What is more, it was obvious to Hutton that both the erosion of the bedrock by rivers, and the accumulation of sediment in the sea, are extremely slow processes. And since no one had witnessed the sort of upheavals in the Earth capable of tilting the layers of older strata from the horizontal to the near vertical, Hutton reasoned that these too must take an immense length of time. Yet the evidence at Siccar Point was that all these events had occurred more than once. The rocks seemed to demand a virtually limitless timespan for their formation. To Hutton and his contemporaries the implications were almost literally staggering. John Playfair, who more than anyone else publicized Hutton's ideas, wrote after he had been shown the unconformity by Hutton:

On us who saw these phenomena for the first time, the impression made will not easily be forgotten…What clearer evidence could we have had of the different formation of these rocks, and of the long interval which separated their formation…Revolutions still more remote appeared in the distance of this extraordinary perspective. The mind seemed to grow giddy by looking so far into the abyss of time.

*Sedimentary rocks, such as these Jurassic sandstones from Utah in North America, are built up of many layers. Each layer is a record of an event in the past – a river in flood or drifting sand in high winds.*

If you visit Siccar Point today, you can still experience the excitement felt by Playfair. With one step, you can move from the rocks above and below the unconformity, traversing aeons of time.

Hutton has had something of a mixed press among historians. He was a curious mixture of the sharp observer, who could see things for what they were, uncluttered by preconceptions, and a theorist with a hidden agenda. He believed that the Earth is a perfect perpetual motion machine, expressly designed by Providence to provide a habitat for us, and he used his field observations to support this idea. According to Hutton, the surface of the Earth is in a constant state of see-sawing: highlands are worn away by rivers and the detritus is carried to the lowlands on the Earth, in the sea, to accumulate; as this detritus accumulates, it compresses under its own weight, slowly heating up. The heated rock expands, uplifting the bed of the sea, so that now the lowlands become high, while the old highlands have been lowered by the action of rivers. And so the process repeats itself in reverse, with rivers flowing from the newly uplifted regions back to the worn-away ones, carrying detritus which will accumulate, heat up and expand yet again, while the existing highlands are eroded once more; and so on, *ad infinitum* (see pp. 18–19).

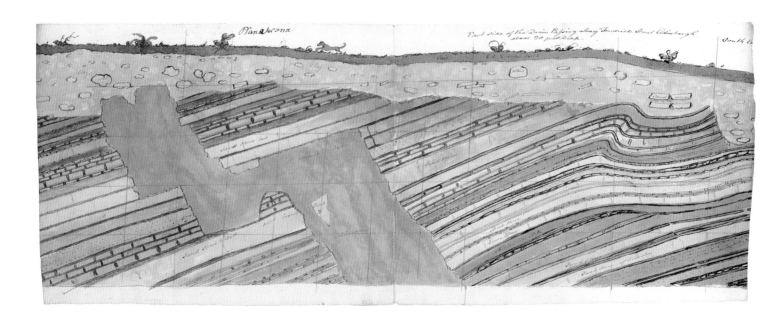

This beautiful watercolour by John Clerk, painted nearly 200 years ago, must be one of the first diagrams to accurately show the nature of rock formations. The rock layers are invaded by a cross-cutting body of once molten igneous rock. An unconformity, where younger horizontal strata rest on top of the older eroded and tilted rock layers, lies just beneath the surface.

Whatever one might feel about the Earth as a perpetual motion machine, or even the role of heat in creating mountains, one cannot deny that Hutton had stumbled on an extremely profound idea: the concept of geological time. In 1788 he wrote of the Earth that one can see 'no vestige of a beginning – no prospect of an end'. This was a radical change indeed, from a concept of Earth history which could be measured in thousands, or even hundreds of thousands, of years, to one which for all practical purposes was infinite. This idea exerted an influence which extended far beyond geologists – poets and writers like Wordsworth and Goethe used the new concept of infinite time as powerful romantic imagery in their works. But Hutton, in adopting a limitless history to the Earth, with no overall direction or sense of evolution, had lost sight of the details of the record in the rocks. It was as though he had not thought it necessary to decipher the individual sentences in the book which he had so painstakingly learned how to read – sentences which told the story of the events which had created the individual layers of rock. The ordering of these layers, and the erection of a timescale in which to place them, were to become the major preoccupations of geologists in the nineteenth century.

## PUTTING A TIMESCALE TO EARTH HISTORY

In 1695, John Woodward, who endowed the first university chair in geology at Cambridge University, had proposed that strange organic-looking shapes which could often be found in rocks were the fossilized remains of those creatures which had failed to get on to Noah's Ark before the great biblical Flood. He imagined that the Flood destroyed and stirred up the landscape and, when the waters subsided, the muddy residue consolidated to form the present bedrock of the planet. Drowned creatures, caught up in the mud, were preserved as fossils.

In the early part of the nineteenth century, Woodward's fossils began to assume an enormous importance to geologists. The first to see their true significance was William Smith, a canal engineer from Oxfordshire. His job involved surveying the route for various canal schemes and supervising the major engineering works involved. He was very much aware of the nature of the bedrock, through which long and deep channels had to be cut. It was clear that the rocks formed a sequence of layers or strata, the characteristics of which, such as colour composition and hardness, he routinely recorded. In the course of his work, which involved extensive travel throughout England, he found that distinctive layers or groups of layers would crop up in widely different locations. Often he could trace these layers for great distances across the landscape. And here he made a breakthrough which must rank with Hutton's discovery of his unconformity as one of the important steps forward in the scientific understanding of our planet.

Ever since he was a boy, Smith had taken an interest in the fossilized creatures that he had found in the rocks – the remains of corals, shellfish, sea urchins and other forms of marine life. He realized that 'each stratum contained organized fossils peculiar to itself'. Examining the fossils became an alternative way of identifying the layers. Smith, probably without much thought, then made an enormous intuitive leap. He assumed that rock layers containing the same distinctive fossils were the *same age*. This way he could classify rock layers not just on the basis of their external characteristics, but also their age. The implications of this step are huge. It provided a way of correlating layers all over the world, merely on the basis of their fossil content.

On the other side of the English Channel, a group of French scientists were making similar discoveries to William Smith at roughly the same time. But unlike Smith, who saw fossils as essentially a tool for recognizing and correlating particular rock layers, these scientists were interested in the detailed features of the fossils themselves. Jean Baptiste Lamarck,

*Rocks are full of the fossilized remains of once-living creatures. Sedimentary layers, deposited at different periods in the past, contain their own unique fossils. For example, this assortment of shells and coiled ammonites lived 150 million years ago in the Jurassic period.*

# The see-sawing landscape

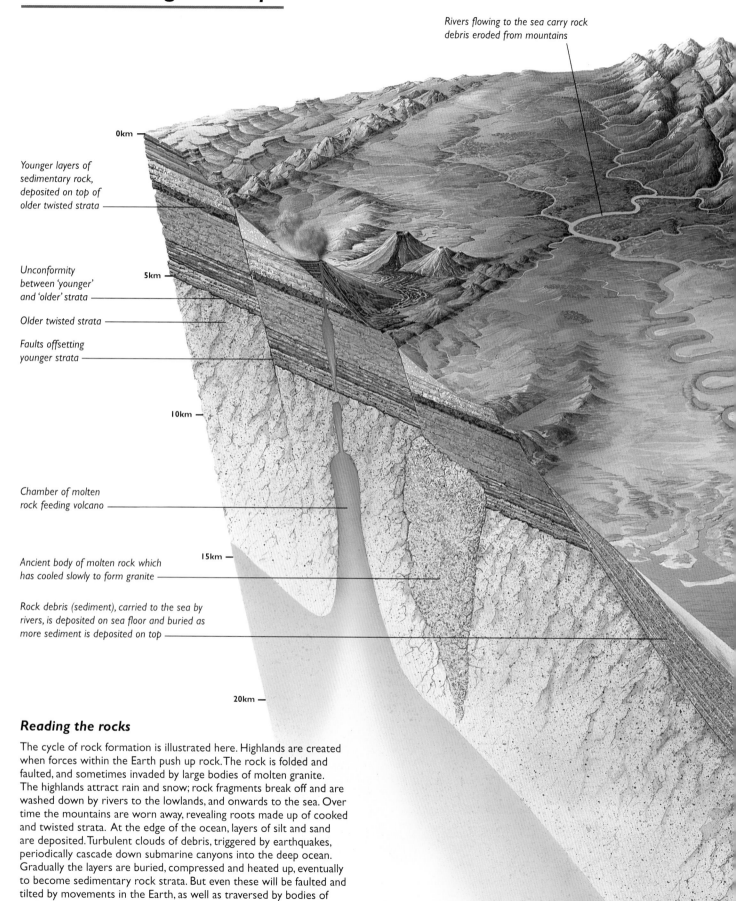

Rivers flowing to the sea carry rock debris eroded from mountains

0km

Younger layers of sedimentary rock, deposited on top of older twisted strata

Unconformity between 'younger' and 'older' strata

5km

Older twisted strata

Faults offsetting younger strata

10km

Chamber of molten rock feeding volcano

15km

Ancient body of molten rock which has cooled slowly to form granite

Rock debris (sediment), carried to the sea by rivers, is deposited on sea floor and buried as more sediment is deposited on top

20km

## Reading the rocks

The cycle of rock formation is illustrated here. Highlands are created when forces within the Earth push up rock. The rock is folded and faulted, and sometimes invaded by large bodies of molten granite. The highlands attract rain and snow; rock fragments break off and are washed down by rivers to the lowlands, and onwards to the sea. Over time the mountains are worn away, revealing roots made up of cooked and twisted strata. At the edge of the ocean, layers of silt and sand are deposited. Turbulent clouds of debris, triggered by earthquakes, periodically cascade down submarine canyons into the deep ocean. Gradually the layers are buried, compressed and heated up, eventually to become sedimentary rock strata. But even these will be faulted and tilted by movements in the Earth, as well as traversed by bodies of molten rock which rise to the surface and erupt in volcanoes. And so the cycle continues, as the landscape see-saws up and down.

Rock layers, folded and broken along faults, are uplifted to form mountain range

Metamorphism – rock heated up by intruding granite body

Edge of deep ocean

Turbulent flows of sediment and water cascade down submarine canyons

*In the early 1800s, Georges Cuvier reconstructed a number of strange beasts from fossilized bones, convincing geologists that many different forms of life had become extinct. This one, called Megatherium, was found in South America.*

Georges Cuvier and Alexandre Brongniart all worked on the rocks of the Paris basin in the early part of the nineteenth century. Cuvier found in the gypsum quarries of Montmartre an animal that was quite unknown, combining the characteristics of a tapir, rhino and pig. He called this creature Palaeotherium ('ancient beast'). From the examination of many fossil bones, Cuvier was able to show that the Earth was once populated by a host of creatures which have since entirely disappeared. It was clear that over geological time many species had gone extinct, while new species had come into existence. Within a few decades, Charles Darwin was to solve the problem of the origin of species with his theory of evolution by natural selection (see Chapter 7).

As geologists started examining the details of the rock record, they also began to recognize evidence for abrupt environmental changes. For instance, an alternating succession of sea creatures and land animals, found in the rock layers near Paris, suggested sudden inundations of the sea in a series of catastrophes. Among those who emphasized the role of sudden upheavals in Earth history was the Reverend William Buckland, who argued that gigantic deluges, like the great biblical Flood, play an important role in past geological events. Buckland has the distinction of giving the first university lecture course in geology at the University of Oxford in 1819. There is a wonderful engraving which shows Buckland, in front of a desk piled high with strange fossils and surrounded by charts and diagrams illustrating the succession of strata in parts of Britain, lecturing to a solemn audience of dons, all dressed in gowns and mortar boards. Buckland endeavoured to show that the geological record is consistent with the biblical Flood. He initially maintained that man had existed, as written in the Bible, since the creation of the planet, but, when no evidence for human remains were found in deposits which he considered to be relics of the Deluge, he relegated the Deluge to before the creation of man. So, little by little, he parted company with the biblical record, and much later in his life accepted that the Earth had a long and complex history before the Flood.

The so-called 'catastrophists' were challenged by Buckland's pupil Charles Lyell, who argued that they had failed to read the rock record properly. Lyell thought that the geological record is inherently incomplete, comprising, as it were, a series of snapshots of the past. Thus, widely separated events are telescoped together, giving the impression of a succession of catastrophes, whereas the true picture is one of slow and steady change. Lyell took this view to its logical conclusion in his classic book *The Principles of Geology*, first published between 1830 and 1833. His ideas rested on two axioms, which together define his 'uniformitarian' philosophy for the history of the Earth: firstly, all processes that can be observed today in the Earth, such as the action of rivers, volcanic activity, or uplift and subsidence of the land, also occurred in the past. In other words, the laws of nature are constant. This has sometimes been summarized by the catchphrase 'the present is the key to the past'. Secondly, the rates at which these processes operate are constant. For instance, changes in sea level, or the wearing away of the landscape by rivers, are slow processes today, and therefore they have always been so in the past.

These are useful principles which still inform the way geologists work. However, Lyell went wrong when he treated them as axiomatic. The guiding principle in modern geology might be that if past events can be explained by a process going on today, then there is no need to invoke an extraordinary explanation or catastrophe. But catastrophes cannot always be ruled out. Geologists today are prepared to consider giant meteorite impacts as an explanation for mass extinctions of life, despite the fact that such giant impacts are not events that happen regularly today.

Nevertheless, by the 1850s the uniformitarian views of geologists like Charles Lyell dominated geological thinking. They were also satisfying to a biologist like Charles Darwin, who recognized that natural selection required a vast amount of time to accomplish the enormous transformation of living things evident from the fossil record. Lyell, like James Hutton before him, was essentially suggesting that there was no overall direction or time limitation to geological processes. But however comfortable scientists had become with this idea, developments in physics were about to throw it into doubt.

*In the last century, industrial processes such as steel milling gave scientists a first-hand experience of the behaviour of hot materials – red- to white-hot steel is at a similar temperature to parts of the Earth's deep interior.*

## A COOLING CALCULATION

The middle part of the nineteenth century was a period of intense industrial development in Britain, powered by coal fires and the steam engine. Engineers and physicists started to develop new concepts of heat and energy, and a new science, thermodynamics, which described their behaviour. The newly discovered laws of thermodynamics set limits to natural processes and ruled out the possibility of perpetual motion, introducing the idea of irreversible change. The notion of processes acting on the Earth for a limitless time at a uniform rate, without a beginning or an end, no longer made any sense. Just as a ticking watch is powered by a spring, the Earth needs a source of energy which, if not replenished, will eventually run out.

In the 1860s, William Thomson, later to become Lord Kelvin, started worrying about the Earth's energy source. He was a physicist at Glasgow University who had carried out pioneering work in thermodynamics

on the transformation of one form of energy into another. Kelvin had been interested in the origin of the heat emitted by the Sun, and concluded that the only possible source was the energy released by the impact of rock bodies as they were drawn into the Sun's mass. According to Kelvin, this energy is the source of the solar radiation and is now being steadily lost into space.

Kelvin was impressed by the fact that as coal miners tunnelled deeper into the Earth they encountered rocks which were hotter than at the surface. This suggested that there was a source of heat within the Earth which was steadily flowing out. Kelvin then postulated that this heat was left over from the time of the planet's formation. If this line of reasoning was correct, Kelvin realized he would be in a position to estimate the age of the Earth. Assuming that initially, the whole surface was molten, then for

*This Brazilian worker is harvesting salt from the sea. Salt is being continually added to the oceans by rivers. A nineteenth-century geologist, John Joly, estimated how long it would take for the world's rivers to make the oceans as salty as they are today – the answer was tens of millions of years.*

all practical purposes the length of Earth history is the time taken for this molten planet to cool to its present state. Kelvin believed he could use the new understanding of thermodynamics to calculate how long this must have taken.

Kelvin's calculation was based on a number of assumptions. Firstly, the only source of heat in the Earth is the primordial heat acquired on its creation through gravitational collapse. Secondly, the Earth has cooled by conduction, in the same way that a hot water bottle cools in a cold bed. Thirdly, the atmosphere of the Earth has always been cool, roughly the same temperature as today. Lastly, he assumed that initially the temperature throughout the Earth was the same as that of molten basalt erupted

from volcanoes (that is, 1100° C). Kelvin also needed to know the thermal properties of the Earth – such as how good it is at retaining heat – and these he derived from measurements on a number of rocks. Finally, he calculated how long a body the size of the Earth, in these circumstances, would take to cool so that the surface temperature gradient would be the same as that observed in coal mines. This calculation would give him the age of the Earth based on sound physical principles – this age, in Kelvin's opinion, would be the maximum length of time that geologists were allowed to juggle with, and it was finite.

The simplicity of Kelvin's argument is both seductive and deceptive. It has the air of mathematical rigour. In fact, many uncertainties exist.

Quantifying the relevant parameters in the Earth, such as the initial temperature, the thermal properties, and the present-day temperature gradients in deep mines, is not easy. In 1862, Kelvin settled on an age somewhere between 20 and 400 million years. By 1897, after he and other physicists had refined the calculations, his estimate narrowed to somewhere between 20 and 40 million years, and most likely nearer 20 than 40 million years.

Kelvin's quantitative approach to Earth history had a profound influence on geologists, inspiring them to their own attempts at gauging geological time. One idea was to measure how long it took for a few layers of sediment to be deposited, and then to scale this up to calculate how much time is needed for the accumulation of the large thickness of strata preserved in the rock record. This method was fraught with problems. The rate of accumulation of sediment is highly variable and is often an intermittent process; also, no one place preserved the complete history of deposition and so the total thickness of sediment laid down since the Earth's formation had to be inferred by building up a rock record from sequences found in different locations. A number of crude estimates of the maximum length of geological time were based on a 10 kilometre thickness of sedimentary rock, and accumulation rates of roughly one metre every 10,000 years. Thus the American palaeontologist Charles Walcott wrote in 1893 that 'geologic time is of great but not indefinite duration. I believe that it can be measured by tens of millions, but not by single millions or hundreds of millions of years'. In 1899, the Irish geologist and physicist John Joly tried another ingenious approach. He tried calculating the time for the oceans to become salty, assuming that they formed early in the Earth's history as fresh water. He imagined that the salt was introduced into the oceans by rivers, and so by estimating the annual load of salt carried by the world's rivers and measuring the saltiness of the sea, he arrived at an age of 99 million years.

By the turn of the century, there was a consensus among geologists that the Earth was about 100 million years old. This figure was a sort of compromise between Lord Kelvin's cooling calculations and their

These bright yellow crystals are an important ore of uranium, called autunite. Uranium is also found in minute quantities in many rocks. Some of this is radioactive, transforming to the element lead at a predictable rate. Geologists can use radioactive decay as a clock to date rocks.

own estimates based on geological considerations. But for some people, 100 million years was an uncomfortably short period of time. Charles Darwin, towards the end of his life, began to lose confidence in his theory of evolution, which he realized could not possibly work on this sort of timescale. He did not live to see a new development in physics which was to resolve the apparent contradictions and remove this major obstacle to his theory.

In 1896 the French physicist Antoine Henri Becquerel noticed that a photographic plate, next to which he had accidentally placed some mineral salts containing uranium, had become blackened. This proved that uranium must somehow give off its own energy. Pierre and Marie Curie showed that thorium had similar properties, as did two new elements, which they named polonium and radium. Marie Curie dubbed the phenomenon radioactivity. In 1902 the New Zealand-born physicist Ernest Rutherford was collaborating with the British chemist Frederick Soddy in Montreal; together they worked out the basic features of the process. They proposed that the atoms of radioactive elements are unstable and can spontaneously 'decay' to a more stable form, usually that of another element, at the same time emitting radiation and heat. It is indicative of the central place that geology held in the minds of scientists that Rutherford immediately realized that this heat generation had implications for Kelvin's cooling calculations.

In 1904, Rutherford visited England and gave talks on the new discovery of radioactivity to packed audiences at both the Royal Society and the Royal Institution. Over 800 people attended the Royal Institution meeting, and it was reported in *The Times* the following day. Lord Kelvin was in the audience, and such was his pre-eminent place in physics that, years later, Rutherford described the meeting in these terms:

> I came into the room, which was half dark, and presently spotted Lord Kelvin in the audience and realized that I was in for trouble at the last part of the speech dealing with the age of the Earth, where my views conflicted with his. To my relief, Kelvin fell fast asleep, but as I came to the important point, I saw the old bird sit up, open an eye and cock a baleful glance at me! Then a sudden inspiration came, and I said Lord Kelvin had limited the age of the Earth, provided no new source of heat was discovered. That prophetic utterance refers to what we are now considering tonight, radium! Behold, the old boy beamed upon me.

To appreciate the significance of this, we should recall that the whole edifice of Kelvin's calculations about the cooling of the Earth was constructed on the premise that the planet had no internal source of energy except the heat left over from its formation. Rutherford had demonstrated a natural process which was generating heat. This was a fatal blow to Kelvin's method. Those who had argued from geological evidence that the Earth must be far older than Kelvin's 20 million years were vindicated. However, many geologists felt they had been bamboozled by arrogant physicists for long enough, and were intensely suspicious. If physicists had got it wrong once, perhaps they had got it wrong again.

As an ironic coda to the Kelvin saga, it turns out that although Kelvin's estimate is indeed grossly wrong, heat generation by radioactive decay is not the main reason why. While it is true that radioactive decay invalidates one of his basic premises, the heat thereby generated in the Earth is not enough to change the cooling time by very much. The fundamental error in Kelvin's work was in fact the assumption that the Earth is cooling mainly by conduction, rather like a hot brick. In Chapter 4, we will describe the evidence that the interior of the Earth is in constant motion and is convecting. Cooling in this situation is a very different physical problem. Interestingly, Kelvin was one of the few nineteenth-century physicists who could have solved such a physical problem. But it was only after Kelvin's death that the information he would have needed became available. By then, scientists had developed new techniques to probe into the Earth's interior.

## A NEW CLOCK

The real significance of the discovery of radioactivity for geologists was that it at last provided a rigorous method to put a timescale to the history of the Earth. By the end of the nineteenth century, geologists were confident about reading the rock record. They knew how to use the field relationships of rock formations to determine the order of geological events, giving relative ages. But they had never been able to determine the absolute age of any event; these

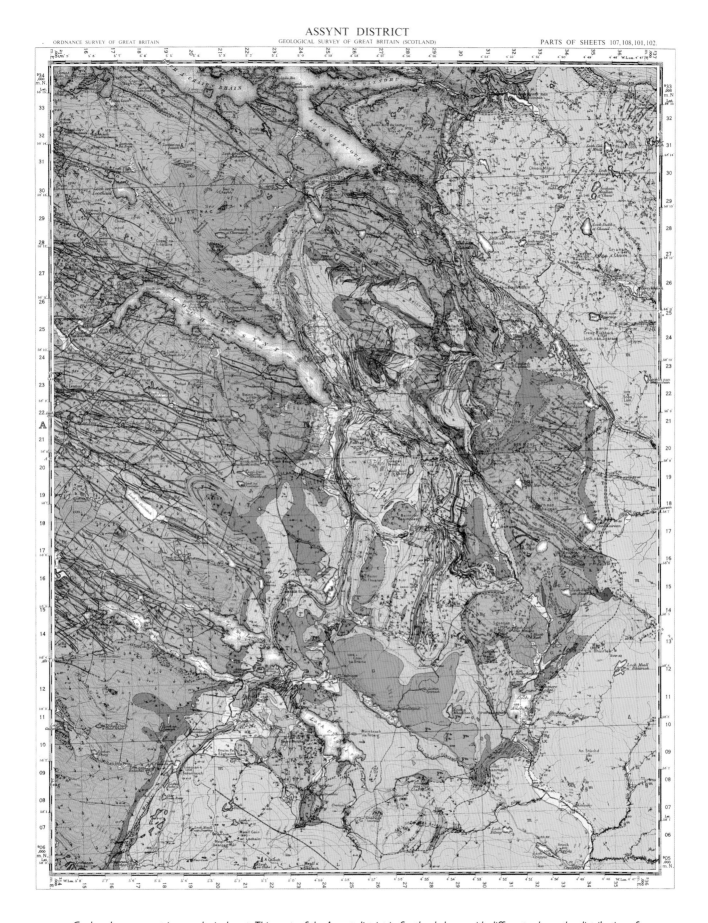

*Geology becomes art in a geological map. This map of the Assynt district in Scotland shows with different colours the distribution of rock formations outcropping at the surface — to make it, geologists spent much time walking over the landscape gathering information and recording the different rock types. Thus the map reveals the story of the rocks.*

# RADIOACTIVE DECAY

**(a)** Granite (rock sample)

Nucleus

Electron orbit →

**(b)** Atom model

Proton

Neutron

Alpha particle emission

Beta (electron) emission

Electromagnetic radiation

**(c)** Nucleus (detail)

Number of parent atoms

1/2

1/4

1/8

1/16

0  1  2  3  4

Leaking bucket

'New' bucket

Number of daughter atoms

15/16

7/8

3/4

1/2

0  1  2  3  4

Time

**(d)** Radioactive exponential decay

Almost all natural rocks contain atoms of radioactive elements **(a–b)**. The nucleus of any of these atoms spontaneously emits particles (alpha or beta particles) and energy, steadily transforming to a new daughter element **(c)**. Geologists can date rocks by comparing the number of daughter and parent elements in the rock today.

The process of decay is rather like a leaking bucket. The initial full bucket represents the number of radioactive parent elements when the rock first formed. An empty bucket underneath catches the drips – this represents the number of daughter elements **(d)**.

In simple situations, such as when a volcanic rock solidifies, there are no daughter elements

to begin with. Gradually, like the water levels in the buckets, the number of daughter elements increases and the number of parent elements decreases. In radioactive decay, the time taken for the number of parent radioactive elements to halve is constant – this is the characteristic half-life of the parent–daughter decay system.

remained stubbornly in the realm of speculation. The key to absolute ages was a method to date rocks directly. In 1905, Rutherford showed that there was a way to do this, using radioactive decay as a natural clock.

Such a clock relies upon the fact that radioactivity is a fundamental atomic process. Unlike almost any other phenomenon which scientists had investigated

up to that time, it is unaffected by such things as temperature, pressure, changes in the chemical environment and so forth. Instead, there is an unvarying probability that an individual radioactive element will decay within a certain time interval. Thus, in a very large population of atoms, the number that will decay in any time interval is proportional to the size of the population. Expressed in another way, this means that for a given original population, the time taken for half these to decay is constant. This time interval is called the half-life, and each radio-active element has a characteristic half-life.

Nuclear physicists have now determined the half-lives of a wide range of radioactive elements, carefully monitoring their decay in nuclear reactors with very sensitive counters. They have found that radioactive decay is generally a very slow process and will carry on over the immense spans of geological time. For most naturally occurring radioactive elements, a half-life of several tens or hundreds of million years is not unusual. Indeed, this is the reason that they have not decayed away to be virtually non-existent but are found in detectable quantities today. When one half-life has elapsed, half the original amount of the element will have decayed away, and half will remain. After two half-lives, half again of what remained will have decayed away, leaving only a quarter of the original number of atoms, and so on. The continuation of this process leads to what is sometimes called exponential decay.

Atoms are the building blocks of matter. They have an internal structure made up of more fundamental particles. These consist of electrons which can be pictured as orbiting a nucleus made up of protons and neutrons. It is the number of protons in the nucleus, called the *atomic number*, which defines the different elements and determines their particular chemical properties. The lightest element, hydrogen, in fact consists only of a single proton with its attendant electron; the nucleus of helium, the next lightest element, contains two protons (and usually two neutrons), and so on up the Periodic Table to uranium, the heaviest element of all, which contains 92 protons. However, the *atomic weight* is determined not by the number of protons alone, but by the sum of the protons and neutrons in the nucleus. The presence of neutrons helps to stabilize the nucleus, but since they do not affect an atom's chemical properties, it is possible for the atoms of an element to have different nuclei with varying numbers of neutrons. Atoms with these different nuclei are called isotopes. Many elements have isotopes which are chemically identical, but have different atomic

weights. For example, helium has two isotopes: helium-4, the nucleus of which contains two protons and two neutrons, and the rarer isotope helium-3, which contains two protons but only a single neutron.

Some isotopes are unstable, and it is these which are radioactive and undergo decay. For instance, there are two radioactive isotopes of uranium, with atomic weights of 235 and 238. These decay to stable isotopes of lead, with atomic weights of 207 and 206 respect-ively. This obviously involves the loss from the nucleus of both protons and neutrons, and this is accomplished through the emission of radiation in the form of alpha particles. Rutherford showed that an alpha particle is in fact an isotope of helium, helium-4 (i.e. two protons and two neutrons). Thus, all uranium-rich rocks will also contain some helium, the amount of which will slowly rise over time as more and more uranium atoms undergo decay (it takes about 700 million years for half the atoms of uranium-235 to decay to lead-207). However, when a rock or mineral is heated to a high temperature, it will lose all the helium trapped inside. So, Rutherford pointed out, the amount of helium gas inside the rock today, which will all have been formed through the decay of radioactive uranium, is a measure of the age of the rock since it was last at a high temperature. This is, in fact, a minimum age, because helium gas is so mobile that some has probably escaped at low temperatures, after the mineral cooled.

In a remarkable experiment which ushered in a new era in geology, Rutherford found that a uranium-rich mineral from Connecticut in the USA must have been in existence for at least 500 million years to account for its concentration of helium. At one stroke he had greatly extended the history of the Earth – such was the power of radioactive dating. Sub-sequently, the Cambridge physicist R. J. Strutt, using the decay of radioactive thorium, which also yields helium, showed that a mineral from Ceylon must have a minimum age of 2400 million years, demonstrating that geological time stretches for billions of years. Later workers used a modified clock, based on the proportions of uranium and lead. Lead is the final decay product of radioactive uranium and, being

*Barberton Greenstone Belt, southern Africa. Here, ancient sequences of rocks, laid down nearly 3.5 billion years ago, provide a window on the early Earth.*

much less mobile than helium, is less likely to be lost from a mineral. When a uranium-rich mineral, such as zircon, first crystallizes during the cooling of molten granite, it does not contain any lead. However, as the uranium decays, lead will accumulate and remain trapped in the mineral. The proportions of uranium and lead, measured today, can be used to calculate the length of time since the granite cooled.

In 1911, the geologist Arthur Holmes began to produce a geological timescale, based on the uranium–lead method. He showed that the part of the geological record which contains shell fossils, called the Phanerozoic by geologists (see Chapter 7), is at least 400 million years long. Rocks from older periods of geological time, referred to as the Precambrian, are billions of years old. In 1927, Holmes estimated the age of the Earth from the relative abundance of radioactive elements and their decay products. He assumed that all the lead in the Earth's crust is created by the decay of uranium and thorium. In this case, the estimates of the abundance of uranium, thorium and lead available to Holmes suggested that the crust was

about 3.6 billion years old. In fact, not all the lead has been created by this type of radioactive decay. Despite this, Holmes's calculation gives some idea of the likely timespan for the history of the Earth. Scientists had moved a long way from the nineteenth-century estimates of a few tens of millions of years.

The technique of radioactive dating has been considerably refined since the pioneering days in the first half of this century. Physicists have accurately determined the half-lives for a whole range of decay systems. They have developed special instruments, called mass spectrometers, which can measure the proportions of chemically identical isotopes extremely precisely. Particular decay series are suited for dating certain types of rocks or geological events. For instance, in addition to the decay of radioactive uranium to lead, the decay of radioactive rubidium to strontium, samarium to neodymium, and potassium to argon, make good geological clocks. Short half-life decay series, such as the decay of carbon-14, can be used to date recent events, including those of human history. This work has shown that the first hominids evolved a few million years ago, the dinosaurs went extinct about 65 million years ago, and the first animals appeared in the fossil record about 600 million years ago. But the geological record, and evidence for life, extends much further back in time. Just as astronomers have revealed that our world is a tiny speck in the immensity of the Universe, so geologists have shown that human history is but the blink of an eye in the vastness of time. The sheer scale of the timespan revealed by modern geology is surely one of the most notable discoveries of science.

## THE OLDEST ROCKS

In the 1960s and 70s, with the development of new isotopic dating techniques, the challenge was to find the oldest rocks on Earth in the hope that they would reveal the early stages in the evolution of our planet. Some remarkable fragments of the surface of the early Earth have been discovered. In southern Africa, a sequence of volcanic and sedimentary rocks was first described in the early part of this century by a mining geologist, A. Hall. These outcrop in the northern Transvaal province near the mining town of Barberton, and also in the kingdom of Swaziland. Barberton had been the site of a gold rush in the 1880s, and the miners referred to the surrounding rocks, because of their distinctive colour, as green-stones. The relation of the greenstones to other rock formations suggested to Hall that they were the oldest rocks so far discovered in South Africa, but he had no idea of their actual age. Today, they are collectively referred to as the Barberton Greenstone Belt. In the 1970s, when geologists started dating the Barberton volcanic rocks and granites, they were surprised to discover that some of them are about 3.5 billion years old. These rocks are truly a window on the early Earth. Similar rocks have now been discovered in the Pilbara region of Western Australia. But there are even older relics of the Earth.

In 1971, Stephen Moorbath, a pioneering geo-chronologist at Oxford University, was studying the rocks on the southwestern coast of Greenland, near Nuuk, which were known to be some of the most ancient yet found. He heard that a mining company prospecting for iron further inland, in the Isua hills,

*Maarten de Wit has been studying the ancient volcanic rocks in the Barberton Greenstone Belt, southern Africa. Here he examines the remains of lavas with distinctive 'pillow' shapes which erupted 3.5 billion years ago on the floor of an ancient ocean.*

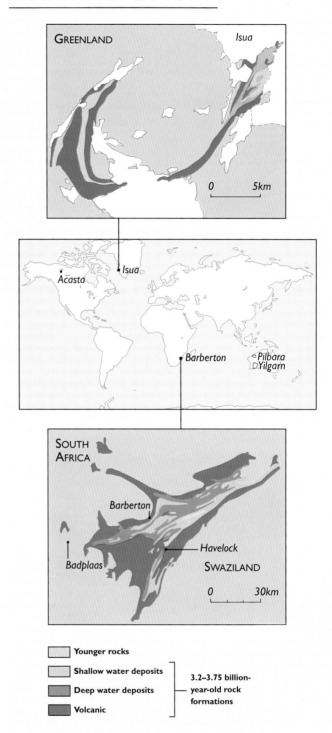

The oldest sedimentary rocks on Earth are found in the continents as small enclaves, surrounded by younger rocks – these provide direct evidence of conditions on the surface of the very early Earth. Sedimentary rocks in the Isua region in Greenland were deposited in a volcanic environment about 3.75 billion years ago. Greenstone belts near Barberton in southern Africa, as well as the Yilgarn and Pilbara regions of western Australia, formed between 3.2 and 3.5 billion years ago. Individual minerals, caught up in younger rocks at Acasta in northern Canada, crystallized during volcanic activity over 4 billion years ago.

was drilling into rocks similar to the ones he was interested in. When he visited this region, he discovered a well-preserved sequence of rocks, consisting of fine-grained sedimentary rocks and lavas. The lavas had characteristic tubular and pillow shapes, like oozing toothpaste, typical of lavas that have erupted under water. Back at Oxford University, Moorbath dated them using the rubidium–strontium method of radioactive dating. This yielded a staggering age of 3.75 billion years. The rocks at Isua are the oldest part of the Earth's surface so far discovered. They seem to be the remains of a volcanic island, where lavas erupted underwater. Among them are finely banded rocks made of iron, brought up in hot springs and precipitated in shallow pools. Even beach pebbles, made up of quartz and volcanic fragments, are locally preserved.

Maarten de Wit, a Dutch geologist, has been working in the Barberton Greenstone Belt for the last twenty years, mapping in great detail these ancient rocks. He, too, has found thick sequences of volcanic rock, with characteristic pillow shapes, which were erupted into water. Some of these volcanic rocks have a much higher concentration of magnesium compared to similar volcanic rocks found today. In fact, there is so much magnesium that geologists have given some of the Barberton volcanic rocks a special name, calling them komatiites after the Komati River, which runs through the Barberton region.

The volcanic nature of komatiites shows that they formed by the melting of the Earth's interior, in a region geologists call the mantle (see Chapter 4). Experiments on producing this type of rock by melting samples of the mantle suggest that the Earth's interior must have been hotter, and richer in water. The heat generated by radioactive decay of the much higher concentration of radioactive elements in the Earth at that time, because less had decayed away, may help to explain why the Earth was hotter. Over billions of years, the Earth has subsequently cooled, and much of the water has been lost into the atmosphere, released during volcanic eruptions.

Within the volcanic sequences, as at Isua, iron-rich sediments are preserved, forming banded rocks called banded iron formation. De Wit has found a distinctive structure in the bands. These look like a series of concentric rings, just like the rings seen today in bubbling mud pools in geothermal regions in New Zealand or South America. To de Wit, this is convincing evidence that the banded iron-rich rock formations are the remains of hot bubbling mud pools. They must have formed when shallow hot mud pools occasionally dried out. There are abundant other signs of powerful geothermal activity, with evidence for

 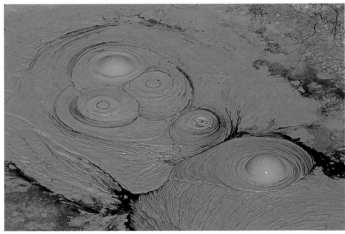

*Tangible evidence of volcanic activity on the surface of the early Earth – these strange concentric structures (left) from the Barberton Greenstone Belt, southern Africa, are the fossilized remains of hot bubbling mud pools which existed on the flanks of a volcano nearly 3.5 billion years ago. They are virtually identical to modern examples (right) from the volcanic region of Bolivia.*

*A vision of the early Earth, billions of years ago, based on the evidence of ancient rock formations. This was a time of intense volcanic activity when primitive life-forms built strange mound-like structures, called stromatolites, in shallow water or thrived near bubbling hot springs.*

explosive geysers which have actually fractured and broken up the surrounding rocks. Here, almost miraculously preserved, are the fossilized remains of primitive bacteria, with their distinctive rice-grain shapes. Clearly, these bacteria were thriving in the hot springs, deriving energy from the volcanic activity.

Overlying the volcanic lavas are the remains of vast ash clouds, which were blasted into the sky after water flowed into the crater and triggered violent eruptions. Thus the oldest Barberton rocks are the remains of volcanoes erupting into shallow seas. Subsequently, sequences of sedimentary rock, several kilometres thick, were deposited between 3.4 and 3.2 billion years ago. There are sandstones, made up of fragments of quartz and granite, which were deposited on beaches, in estuaries and by rivers. Some of these sandstones contain the signs of tidal currents – small sand dunes which migrate back and forth as the tide turns. It is clear that the Earth–Moon system existed at this time. The composition of the sandstones, made up of fragments of quartz and granite, tells us something more. These sandstones

were most probably made up of rock eroded from a region which was unlike the oceans – a region of the Earth's surface where there were huge amounts of granite. Today, such regions are in the continents. Thus, we may be witnessing in the Barberton rocks the formation of the very early continents on Earth.

The rocks of Isua and Barberton really mark the end of the road for geologists on their journey back through time. Often armed with nothing more than a geological hammer, geologists have managed to produce a series of pictures of the Earth at stages in its history so far back in time that we can barely comprehend the passage of the aeons. They now know that the Earth came into existence about 4.55 billion years ago during the creation of the Solar System – a topic for a later chapter (see Chapter 8) – so their journey on Earth has revealed nearly nine-tenths of our planet's history. But it has revealed something more – the workings of our planet. The Earth may have changed, but the forces that shape it have not. In the next few chapters we explore this discovery.

# CHAPTER  2

# THE DEEP

·····································

A curious feature of our planet's surface is that it has
two distinct levels: the dry land on the continents, on
average a few hundred metres above sea level, and
the ocean floor, making up two-thirds of the Earth's
surface, several kilometres below sea level. Only in
the past fifty years have scientists begun to explore
in detail this vast region, revealing beneath the waves
a landscape quite unlike the world we are used to.
They have discovered a vast mountain range which
encircles the entire globe. Here new sea floor is
being continuously formed as the Earth's
surface splits apart.

*The USS Alvin submersible, made of titanium alloy and designed to withstand pressures at depths of
several kilometres in the oceans. Geologists have used the Alvin to explore the mid-ocean ridges.*

## A GIGANTIC JIGSAW PUZZLE

In 1965, Sir Edward Bullard at Cambridge University enlisted the aid of a computer to solve a particularly awkward jigsaw puzzle. He was interested to see if it was possible to find a fit between the western margin of Africa and the eastern margin of South America. The computer tested numerous ways of placing the two continents together, searching for the one that minimized overlaps or gaps. Bullard found that if Brazil is tucked into the bend between West and Central Africa, the fit is, statistically speaking, extraordinarily good. Slight overlaps exist, such as where the Niger River flows into the South Atlantic, but the mismatches are very minor. Bullard went on to use the same technique to fit North America with Greenland and Europe. Again the fit was good, with Greenland's eastern margin resting snugly against Scandinavia and Great Britain, and its western margin pushed up against northern Canada. But in making these fits, Bullard had to do something remarkable to the surface of the Earth – he had to remove the Atlantic.

Bullard's geometrical exercise was the culmination of over fifty years of speculation by geologists about the movement of the continents. Geologists had ranged themselves into two opposed camps: the 'fixists' who believed that continents have always been where they are today, and the 'mobilists' who believed that continents drift over the surface of the Earth. It is easy to see why the latter were often considered slightly cranky. Who in their right mind would suggest that continents move around? Certainly, the most eminent geophysicist of the first half of the twentieth century, Sir Harold Jeffreys, thought that it was a crazy idea. His observations of the way the Earth responds to earthquakes suggested to him that the interior of the Earth was far too rigid – in fact as rigid as steel – to allow the surface to move around. The mobilists had not managed to come up with a good answer to this objection.

The driving force behind the mobilists was a German scientist called Alfred Wegener. Wegener was born in 1880, and as a young man he was fascinated by both the new science of meteorology and the exploration of Greenland. It is said that while in Greenland he was inspired by the motion of floating ice to conceive of the idea of drifting continents: certainly, ice floes in water, as they break up and drift apart, are very close to Wegener's concept of continents on the surface of the Earth. Once the idea took hold, he began to marshal all the evidence he could find to support it, producing a book in 1915 with the grand title *Die Entstehung der Kontinente und Ozeane* (*The Origin of Continents and Oceans*). Wegener proposed that in the past, all the continents were joined together to form one enormous supercontinent which he called Pangea (Greek for 'all land'). Subsequently, this supercontinent broke up and the fragments drifted apart to their present positions. The gaps which opened up between the fragments became the present oceans.

There were two main lines of evidence that Wegener relied on to support his idea. The first was that he could match features on one continent with those on another. In Wegener's own words: 'it is just as if we were to refit the torn pieces of a newspaper by matching their edges and then check whether the lines of print run smoothly across'. The 'lines of print' were found in the rock formations on widely separated continents: evidence either for great similarities in ancient faunas and floras which could not conceivably have been in contact across a wide ocean, or for identical geological deposits of similar age on different continents.

Wegener showed that if the continents are put together to assemble the 'original' continent of Pangea, then the distribution of ancient floras, faunas, ice sheets, deserts and temperate zones, indicated by the geological record, fit together. For instance, the fossilized remains of the Glossopteris seed fern, which flourished about 270 million years ago, is found today on the widely separated continents of South America, southern Africa, India and Australia. In Wegener's reconstruction of Pangea, this seed fern lived in a single vast region. Similarly, scattered signs of an ancient Ice Age in South America and Africa, several hundred million years ago, would have been left

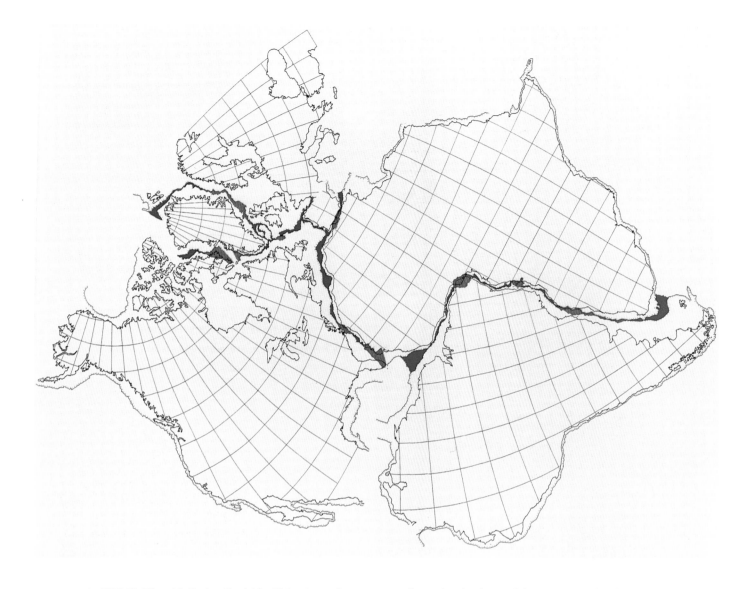

In 1965, Sir Edward Bullard at Cambridge University used a computer to fit together the shapes of the continents with only a tiny area of mismatch, shown by the blue and red regions. This strongly supports the notion that the continents were once joined together and have subsequently rifted apart to form the Atlantic Ocean.

behind by a single continental ice sheet near the south pole.

Wegener thought that the continents are still moving apart today. He was particularly impressed with the results of a surveying expedition to Greenland in 1906 which suggested that Greenland had moved westward relative to Denmark several hundred metres since a previous survey in 1870. This remarkable result seemed to prove the reality of continental drift. Unfortunately it was wrong, based on faulty surveying. When the measurements were repeated several times between 1927 and 1948, such a large movement could not be substantiated. Wegener had died in 1930, and the failure of what he himself considered to be a crucial test of his idea, combined with other flawed supporting data and the lack of a plausible driving mechanism, ultimately destroyed his credibility.

By the Second World War, most leading geo-physicists did not take the notion of drifting continents seriously. But the arguments for and against continental drift had been made from the perspective of the land, while Wegener's ideas also required the creation of new oceans between the fragments of Pangea. Nobody knew much about this part of the Earth. But, as that began to change in the first decade after the Second World War, so did the credibility of Wegener's idea.

## VOYAGE INTO THE UNKNOWN

The 1950s marked the height of the Cold War. The rivalry between the United States of America and the Soviet Union was turning the seas into a battleground for a new type of weapon – the long-range nuclear

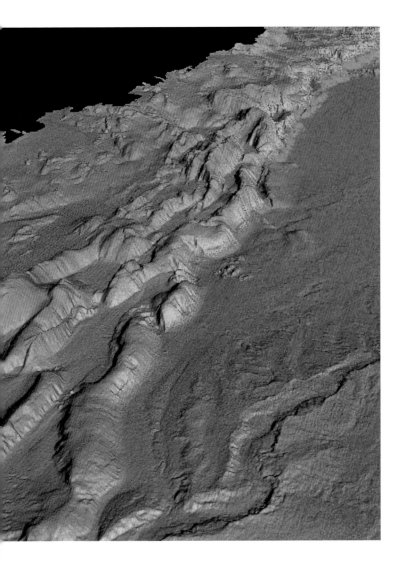

*Scientists today have a detailed picture of the shape of the sea floor. This high definition sonar image of the Pacific Ocean off the coast of Oregon, western North America, looking south, shows the shallow continental shelf (coloured pink–yellow), a few hundred metres deep, and the much deeper ocean floor, over 3 kilometres deep, further offshore (coloured blue–green).*

submarine. The US Navy suddenly needed to know about the world's oceans. They wanted accurate charts of the sea floor to help with submarine navigation. They wanted maps of the varying gravitational field over the oceans to calculate nuclear missile trajectories. They wanted studies of Pacific Ocean islands because these made isolated sites to test atomic bombs. And they did not seem to mind how much money they spent. For scientists studying the oceans it was a bonanza: between 1948 and 1958 the US Government increased its funding of academic ocean research nearly ten-fold.

Science at sea was a hazardous business. Oceanographers worked constantly with explosives and heavy equipment on small ships in heavy seas. There were numerous accidents – scientists were swept overboard or crushed by loose bits of machinery. Weeks or months at sea told heavily on people's married lives as well. But the 'sea scientists' were almost addicted to this activity, and a small clique virtually dominated the research. The names of Maurice Ewing, Bruce Heezen and Bill Menard crop up again and again in the important scientific publications of the 1950s.

The decade began with a research cruise in the Pacific Ocean, operating out of the Scripps Oceanographic Institute on the west coast of the United States. The MidPac expedition, as this voyage was called, was destined to confound many people's expectations. Scientists at the time had a mental picture of much of the very deep sea floor as a vast smooth plain, occasionally broken by islands and submarine mounds, with gentle warps or rises. Much of this picture had come from the work of the scientists aboard HMS *Challenger* in the early 1870s. They had discovered the general shape of the sea floor by laboriously taking soundings – lowering a lead weight attached to a length of piano wire over the side of the ship.

The *Challenger* expedition showed that the deep sea does not start immediately off shore. All round the margins of the continents, a region of relatively shallow water, up to 200 metres deep, forms the continental shelf, extending a few hundred kilometres from the coast. Seaward of the shelf, the water deepens to several thousand metres. The oceans are so wide compared to their depth that there is a tendency to produce maps of the sea floor with enormous vertical exaggeration. On such maps, the edge of the continental shelf looks like a giant underwater cliff-face. In fact the sea floor slopes, on average, fairly gently everywhere, but nonetheless there are marked changes in depth. The most important of these are broad zones of shallowing called the mid-ocean ridges, which are really underwater mountain ranges. For example, a mountain range, about 1000 kilometres wide and up to 2500 metres high, runs right down the middle of the Atlantic. In some oceans, usually near the margins, there are so-called trenches, deep narrow depressions about 100 kilometres wide. These trenches are the lowest parts of the Earth's surface, reaching depths of 11 kilometres below sea level off the Mariana Islands in the western Pacific.

Oceanographers had revealed the shape of the ocean floor. But their geological co-workers were interested in rocks – they wanted to know what the floor of the ocean was made of. Many geologists had assumed that the rock beneath the oceans must be similar to that found on the continents, which basically consists of thick piles of ancient sedimentary rock, often cooked up and twisted and invaded by large bodies of once molten granite. Radioactive dating had shown that parts of the continents were billions of years old. So, the oceans were probably old features of the planet as well. It even seemed possible that the Pacific was created when a large fragment of the surface of the Earth splintered off billions of years ago to form the Moon. Over the aeons, so the reasoning went, detritus eroded from the surrounding continents would have been washed into the oceans, blanketing the bottom with a smooth layer of sediment which was expected to be about 5 kilometres thick.

The MidPac expedition had on board a new high-precision instrument called a continuously recording echo-sounder. This was the first time such a device had been used extensively in the deep ocean. It was first developed during the Second World War to detect submarines by emitting pulses of sound at regular intervals. These travel from the bottom of the ship to

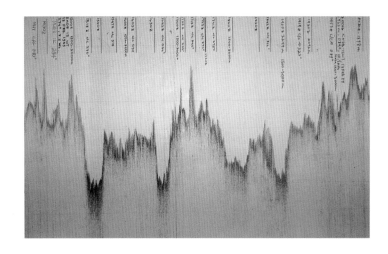

*Scientists routinely use an echo-sounder to measure the depth of the sea floor along a ship's track. This record reveals the rugged topography (greatly exaggerated) of the sea floor near the crest of a mid-ocean ridge.*

the sea floor (or enemy submarine), and then bounce back to be detected at the ship again. The water depth can be calculated from the time delay between the emitted and reflected sound signal, using the known speed of sound in water. The depth is plotted continuously on a chart recorder, producing a picture of the sea floor as the ship sails along (see above). A dangerously close submarine would appear on the chart as a sudden shallowing of the ocean. The echo-sounder revealed that much of the Pacific Ocean floor was surprisingly hilly, quite unlike the bottom of an ancient basin, which would be smoothed out by a thick blanket of sediment from the continents. This was the first evidence to suggest that the ocean floor was not as old as was generally assumed.

A much more powerful type of echo-sounder was used to probe beneath the sea floor. By letting off explosions in the water, the MidPac scientists managed to generate vibrations which penetrated deep into the underlying rocks before bouncing back. The return signal was detected by a series of sensitive microphones towed behind the ship. It is possible, by measuring the time delay between the explosion and the arrival of vibrations at several widely spaced microphones, to calculate the speed of the vibrations. This speed depends on the type of rock through

# The ocean crust

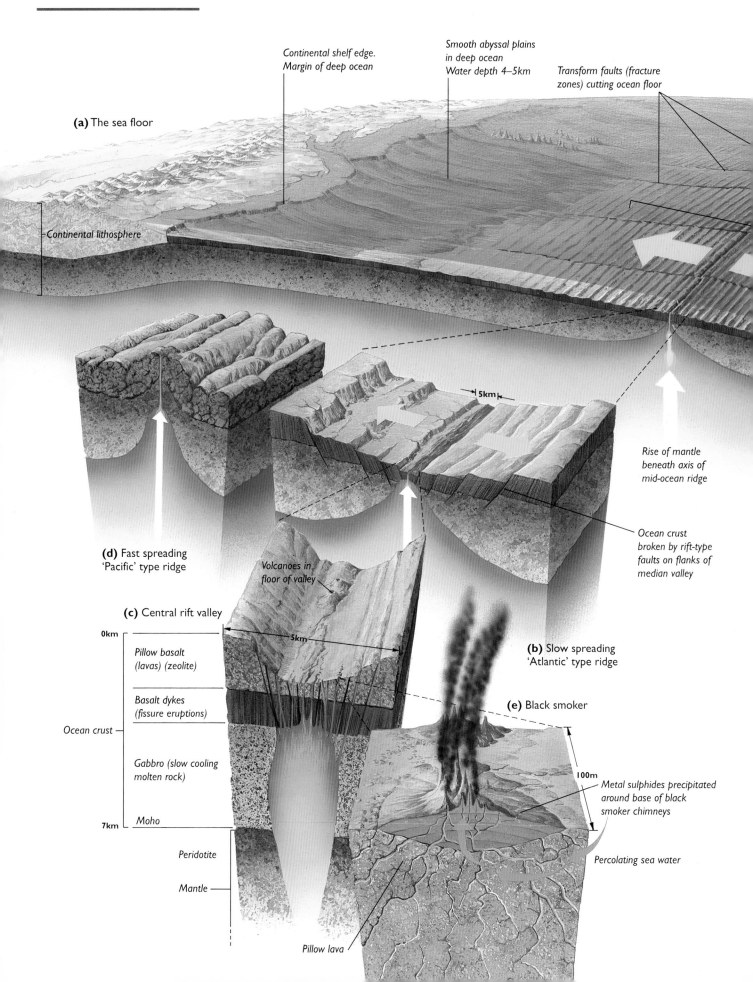

**(a)** The sea floor

Continental shelf edge.
Margin of deep ocean

Smooth abyssal plains
in deep ocean
Water depth 4–5km

Transform faults (fracture
zones) cutting ocean floor

Continental lithosphere

Rise of mantle
beneath axis of
mid-ocean ridge

Ocean crust
broken by rift-type
faults on flanks of
median valley

5km

**(d)** Fast spreading
'Pacific' type ridge

Volcanoes in
floor of valley

**(c)** Central rift valley

5km

**(b)** Slow spreading
'Atlantic' type ridge

0km

Pillow basalt
(lavas) (zeolite)

Basalt dykes
(fissure eruptions)

**(e)** Black smoker

Ocean crust

Gabbro (slow cooling
molten rock)

100m

Metal sulphides precipitated
around base of black
smoker chimneys

7km    Moho

Peridotite

Percolating sea water

Mantle

Pillow lava

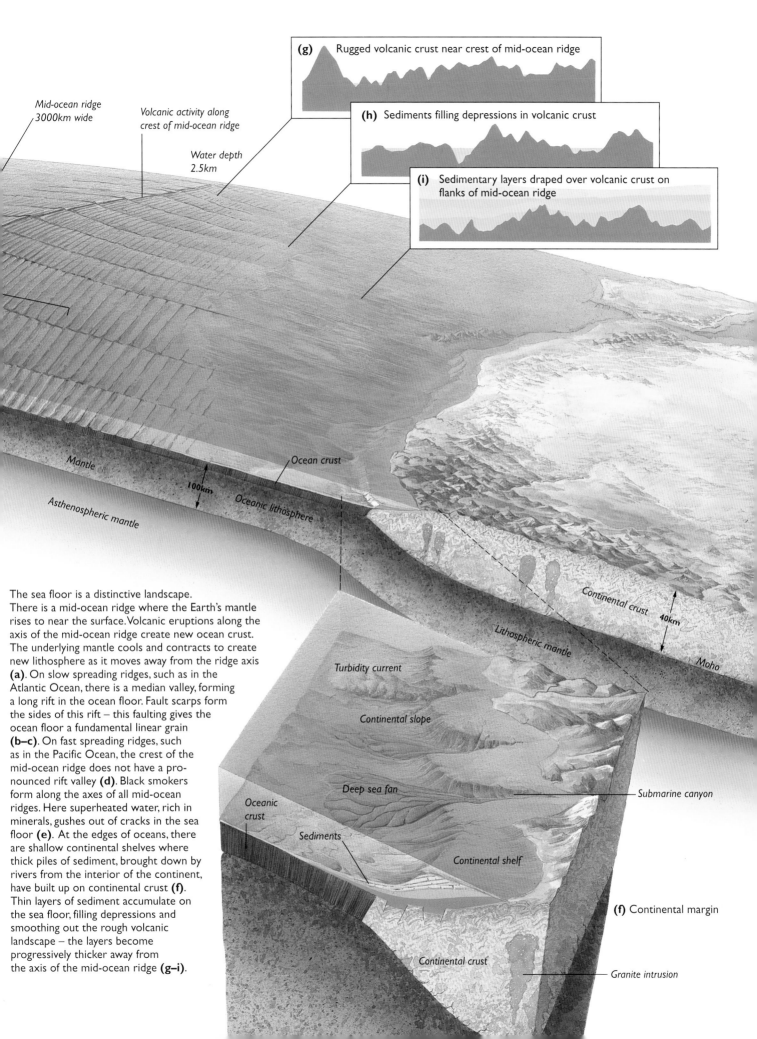

**(g)** Rugged volcanic crust near crest of mid-ocean ridge

**(h)** Sediments filling depressions in volcanic crust

**(i)** Sedimentary layers draped over volcanic crust on flanks of mid-ocean ridge

Mid-ocean ridge 3000km wide

Volcanic activity along crest of mid-ocean ridge

Water depth 2.5km

Mantle

Asthenospheric mantle

Ocean crust

100km

Oceanic lithosphere

Continental crust

40km

Lithospheric mantle

Moho

Turbidity current

Continental slope

Deep sea fan

Submarine canyon

Oceanic crust

Sediments

Continental shelf

**(f)** Continental margin

Continental crust

Granite intrusion

The sea floor is a distinctive landscape. There is a mid-ocean ridge where the Earth's mantle rises to near the surface. Volcanic eruptions along the axis of the mid-ocean ridge create new ocean crust. The underlying mantle cools and contracts to create new lithosphere as it moves away from the ridge axis **(a)**. On slow spreading ridges, such as in the Atlantic Ocean, there is a median valley, forming a long rift in the ocean floor. Fault scarps form the sides of this rift – this faulting gives the ocean floor a fundamental linear grain **(b–c)**. On fast spreading ridges, such as in the Pacific Ocean, the crest of the mid-ocean ridge does not have a pronounced rift valley **(d)**. Black smokers form along the axes of all mid-ocean ridges. Here superheated water, rich in minerals, gushes out of cracks in the sea floor **(e)**. At the edges of oceans, there are shallow continental shelves where thick piles of sediment, brought down by rivers from the interior of the continent, have built up on continental crust **(f)**. Thin layers of sediment accumulate on the sea floor, filling depressions and smoothing out the rough volcanic landscape – the layers become progressively thicker away from the axis of the mid-ocean ridge **(g–i)**.

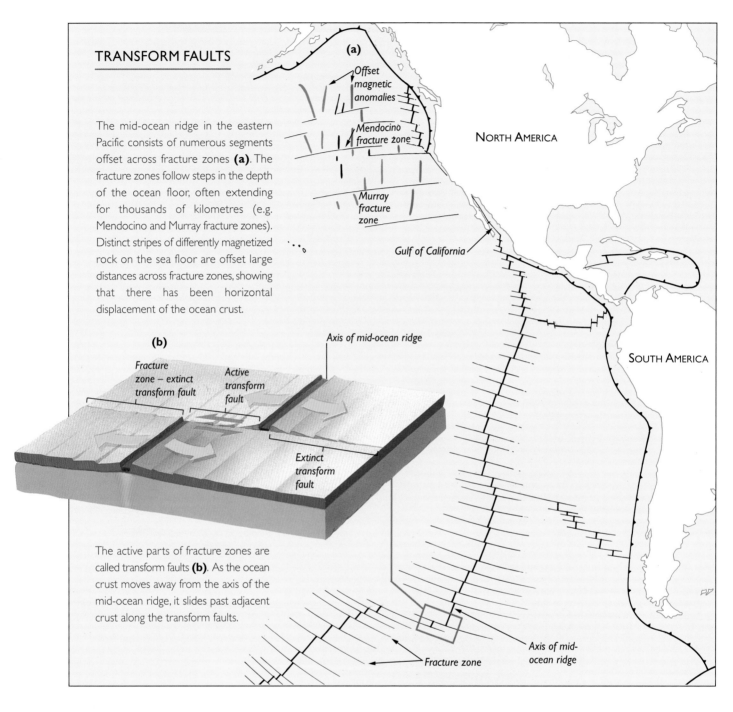

## TRANSFORM FAULTS

The mid-ocean ridge in the eastern Pacific consists of numerous segments offset across fracture zones **(a)**. The fracture zones follow steps in the depth of the ocean floor, often extending for thousands of kilometres (e.g. Mendocino and Murray fracture zones). Distinct stripes of differently magnetized rock on the sea floor are offset large distances across fracture zones, showing that there has been horizontal displacement of the ocean crust.

The active parts of fracture zones are called transform faults **(b)**. As the ocean crust moves away from the axis of the mid-ocean ridge, it slides past adjacent crust along the transform faults.

(a)
Offset magnetic anomalies
Mendocino fracture zone
Murray fracture zone
NORTH AMERICA
Gulf of California
SOUTH AMERICA

(b)
Fracture zone – extinct transform fault
Active transform fault
Axis of mid-ocean ridge
Extinct transform fault
Fracture zone
Axis of mid-ocean ridge

which they have passed. A similar technique had been used on land to reveal the basic structure of the continents. Here, vibrations travelled at character-istically slow speeds down to a depth of roughly 35 kilometres. Below this, the speed of vibrations was significantly faster. Geologists had dubbed the layer through which vibrations travel relatively slowly the crust. The deeper layer was called the mantle. The

MidPac scientists found both a crustal and a mantle layer beneath the Pacific Ocean. But strangely, the ocean crust was much thinner than the crust beneath the continents, with a remarkably constant thickness of about 7 kilometres.

The MidPac scientists managed to tease out a more detailed picture of the ocean crust from their seismic probing. The top part of the crust was a layer only about a kilometre thick in which vibrations travelled the slowest. The scientists guessed that this could be sedimentary cover, but much thinner than expected. At this stage, nobody knew the composition of the rest of the crust – this was an important mystery which was solved only many years later. But the MidPac team did finally manage to get hold of some rock from the sea floor. After many attempts, dragging a long cable with a metal scoop at one end, they brought up from depths of several kilometres isolated fragments of fossilized coral. Palaeontologists immediately recognized that these had once flourished in shallow water as part of a coral reef about 100 million years ago. Since then, the ocean floor must have subsided to its present great depth. This was another hint that the deep oceans might be relatively young features of the Earth's surface – much younger than the billions of years many scientists were predicting. Subsequent voyages in the early 1950s showed that the results of the MidPac expedition were not atypical – large parts of the floor of the Pacific and Atlantic Oceans share similar features, with a crust much thinner than that in the continents.

Bill Menard, who was one of the American scientists on board the MidPac cruise, made another unexpected discovery on a later voyage in the Pacific. As a naval employee, he had access to US Navy charts of this region. He noticed abrupt changes in the depth of the deep sea floor which appeared to occur across roughly east–west lines. These changes in depth are almost at right angles to the general 'grain' of the sea floor, which is marked by roughly north-trending hills and valleys. Menard soon had a chance to investigate this when he joined a research cruise heading to the Pacific island of Midway. Using an echo-sounder, he detected a step-like feature in the sea floor about 2000 metres high. Zigzagging across this step, it became clear it extends for over a thousand kilometres in an east–west direction. Menard had discovered a new type of undersea mountain range. By 1953, he had discovered several of these, of which the longest extends for nearly 5000 kilometres in the eastern Pacific. The top of each step is a vast slab of deep sea floor which tilts gently away from the edge of the step. Often the height of the step

progressively decreases in one direction. Menard called all these features fracture zones, because on a large scale they look like gigantic breaks or fractures in the sea floor (see opposite).

In 1956, Bruce C. Heezen and his colleague Marie Tharp, based on the other side of the United States to the MidPac scientists, at the Lamont-Doherty Observatory outside New York, used the huge quantity of ocean depth soundings to produce the first detailed map of the topography of the world's sea floor. The map (overleaf) shows oceans which are generally about 4 kilometres deep, but tend to be shallower towards the centre, forming a mid-ocean ridge. The summit of the mid-ocean ridge is about 2.5 kilometres below sea level. Most of the ocean floor is quite rough with a distinct grain parallel to the mid-ocean ridge, made up of numerous elongate hills. Only near the foot of the continental shelf escarpment is the deep ocean floor really smooth, in regions called abyssal plains. Here, it is blanketed by a thick cover of sediments, brought into the oceans by the rivers.

In making their map, Heezen and Tharp stumbled on an extraordinary fact: the mid-ocean ridge forms a virtually continuous undersea mountain range which snakes right round the planet. In fact, this is the longest mountain range on Earth. It has a number of unusual features. For much of its length, a valley runs right down the centre, looking very much like the system of valleys in East Africa where the Earth's crust is splitting apart. Bill Menard's fracture zones form another system of linear features which run nearly at right angles to the mid-ocean ridge. To begin with, the fracture zone mountains were often confused with the mid-ocean ridge mountains, but by 1958 it was becoming clear that the mid-ocean ridges were broken up into discrete segments between fracture zones.

*Overleaf: In the late 1950s, the American oceanographers Bruce C. Heezen and Marie Tharp used all the available soundings of the world's oceans to produce the first detailed picture of the sea floor – the mid-ocean ridges form a gigantic chain of undersea mountains which circle the world. This version of their map of the World Ocean Floor was published in 1977.*

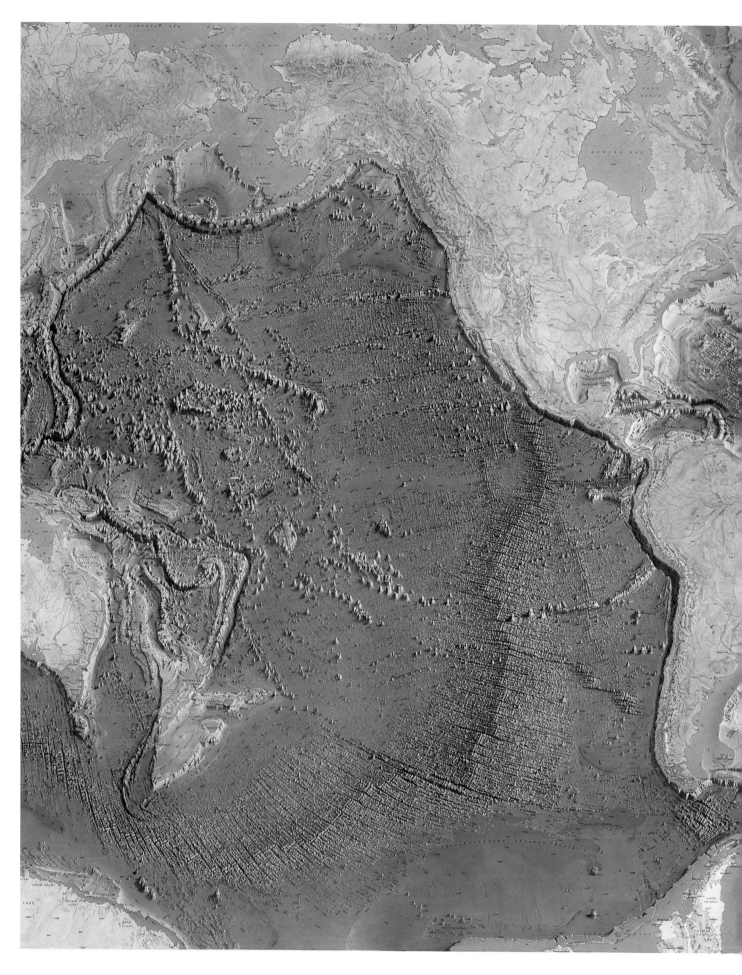

WORLD OCEAN FLOOR

*BY BRUCE C. HEEZEN AND MARIE THARP*

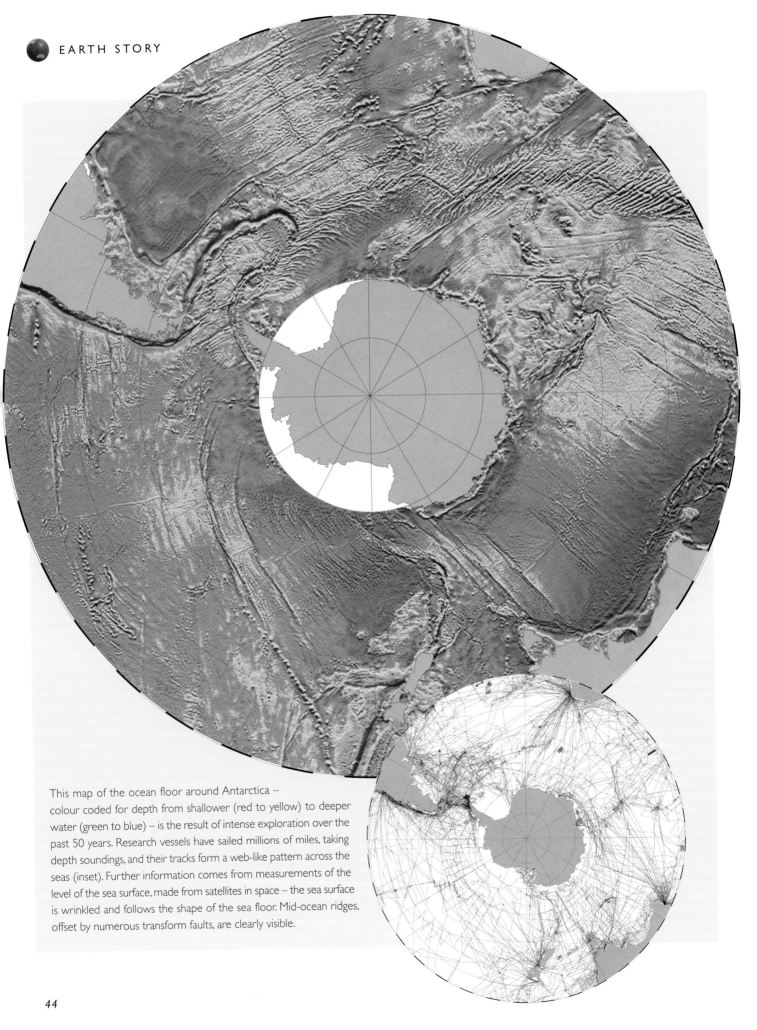

This map of the ocean floor around Antarctica – colour coded for depth from shallower (red to yellow) to deeper water (green to blue) – is the result of intense exploration over the past 50 years. Research vessels have sailed millions of miles, taking depth soundings, and their tracks form a web-like pattern across the seas (inset). Further information comes from measurements of the level of the sea surface, made from satellites in space – the sea surface is wrinkled and follows the shape of the sea floor. Mid-ocean ridges, offset by numerous transform faults, are clearly visible.

## A MAGNETIC ATTRACTION

As the Cold War intensified, the US Navy became concerned about Russian submarines prowling close to US harbours. In the early 1950s, they had installed microphones on the shallow sea bed to listen out for the engines of submarines. But they found that the listening system could be confused by reverberations from underwater hills. The Navy badly needed an extremely accurate map of the depth of the sea floor along the west coast of North America, extending about 600 kilometres off-shore. In 1955, the US Coast and Geodetic Survey was contracted to produce the first really high resolution survey of the sea floor. A ship would sail back and forth on closely spaced parallel tracks, taking soundings. The Navy offered to carry out any additional surveys for research scientists

which did not interfere with the main task. So, it was agreed that an instrument designed to measure the strength of the Earth's magnetic field would be towed behind the ship. This seemed to be the only type of research that would not hold up the Navy's work, but was still potentially capable of telling geologists something new. The idea was that some rocks are slightly magnetic, so that variations in the magnetic field in the oceans might reveal something about the nature of the rocks on the ocean floor. The scientists, however, had no idea what they would find.

The magnetic survey off the west coast of North America was supervised by Arthur Raff. He used a new device called a proton magnetometer, which is capable of measuring the strength of the magnetic field extremely precisely. The Earth's magnetic field does not have the same strength everywhere. Scientists knew already about an overall smooth decrease in the strength of the field from the magnetic north and south poles to the equator, which could be

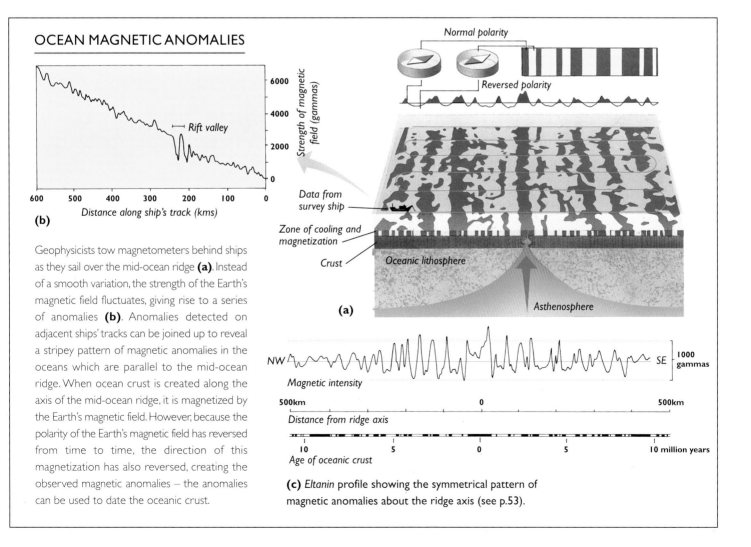

### OCEAN MAGNETIC ANOMALIES

Geophysicists tow magnetometers behind ships as they sail over the mid-ocean ridge (a). Instead of a smooth variation, the strength of the Earth's magnetic field fluctuates, giving rise to a series of anomalies (b). Anomalies detected on adjacent ships' tracks can be joined up to reveal a stripey pattern of magnetic anomalies in the oceans which are parallel to the mid-ocean ridge. When ocean crust is created along the axis of the mid-ocean ridge, it is magnetized by the Earth's magnetic field. However, because the polarity of the Earth's magnetic field has reversed from time to time, the direction of this magnetization has also reversed, creating the observed magnetic anomalies – the anomalies can be used to date the oceanic crust.

(c) *Eltanin* profile showing the symmetrical pattern of magnetic anomalies about the ridge axis (see p.53).

*Measurements of the strength of the Earth's magnetic field reveal a complex pattern of highs (red to brown colours) and lows (green to blue colours) called magnetic anomalies. This map shows the magnetic anomalies along the Oregon coastline of North America. On land (right-hand side of map), the magnetic anomalies form an irregular and blobby pattern. By contrast, the magnetic anomalies in the ocean (left-hand side of map) consist of very regular stripes. The stripes are parallel to the mid-ocean ridge, though occasionally offset by fracture zones.*

predicted fairly accurately. They were looking for smaller-scale wobbles in the field strength. These wobbles are called magnetic anomalies because they deviate from the expected smooth variation. Raff meticulously plotted the magnetic anomalies, as the ship sailed up and down on its parallel tracks. He found that these produced a wave-like pattern of lows and highs which were less than 10 per cent of the total field strength. It became clear that individual magnetic anomalies, measured along one ship's track, could be recognized on nearby tracks. As more and more of these correlations were made, Raff linked up the anomalies from one track to another, and a map of magnetic anomalies in the ocean floor emerged. If

the highs and lows in the magnetic anomaly are marked by white and black, respectively, on the map, then the magnetic map of the ocean floor looks a bit like the markings of a zebra, consisting of parallel stripes a few kilometres to a few tens of kilometres wide. These are quite unlike anything observed on land, where anomalies tend to form blobs of high or low magnetic strength, depending on the irregular distribution of magnetic rocks.

Raff's parallel stripes were so extraordinary that many of his colleagues thought they were the result of some instrumental malfunction, but Raff believed they were real. He also noticed that the anomalies were parallel to the mid-ocean ridge. This observation proved to be of crucial significance, but in 1958, when the magnetic anomaly map was finally published, it was just a curiosity. However, it was immediately clear from the map that the pattern of magnetic stripes was very distinctive. Identical sequences of stripes could be recognized either side of the giant fracture zones in the eastern Pacific, but in some cases these sequences were displaced over 1000 kilometres across the fracture zones. When Bill Menard heard about this, he realized that this observation supported his earlier suspicion that the fracture zones are faults in the ocean floor which have moved sideways.

In the late 1950s, the most fundamental question facing anybody who examined the magnetic stripes was their origin. Magnetic anomalies on land are usually a consequence of variations in rock type. When more magnetic rocks occur near the surface, then a 'high' in the magnetic strength is often observed. But it seemed implausible to think that the whole ocean floor, occupying nearly two-thirds of the Earth's surface, could be made up of alternating stripes of different rock types. In fact, all the evidence indicated that the rocks of the ocean floor are very uniform, with a roughly constant thickness.

## NORTH IS SOUTH

In the 1920s, a Japanese scientist called Motonari Matuyama was studying the origin of magnetism in rocks. It was well known that many rocks are permanently magnetic, like a bar magnet, because they contain iron-rich minerals such as magnetite and haematite. The magnetism in these minerals has a peculiar property. For example, if magnetite is heated above a critical temperature, known as the Curie temperature (about $500°C$), it will lose its permanent magnetism. If it is then cooled down again below this temperature in a magnetic field, it is magnetized again and retains a memory of the surrounding magnetic field. This new magnetization is called a remanent magnetism. In this way, lava or other volcanic rocks which contain magnetic minerals acquire a remanent

magnetism. This magnetism is like that of a simple bar magnet, such as a compass needle, and will tend to orientate itself so as to point towards the Earth's magnetic north pole.

Matuyama analysed in his laboratory at Kyoto Imperial University some volcanic rocks collected in Japan. Naturally, he expected the remanent magnetism to point towards the Earth's magnetic north pole. But he found that in some of his samples it actually points to the south pole, in exactly the wrong direction. Volcanoes with a south-pointing remanent magnetism were older than those with the normal north-pointing direction. The simplest explanation seemed to be that the north and south poles of the Earth's magnetic field had switched a few million years ago. Thus, the older volcanic rocks had acquired a south-pointing remanent magnetism because they had cooled when the magnetic north was actually at the south pole. Interestingly, this radical idea was not dismissed out of hand by his colleagues when Matuyama published his results in 1929. Nobody knew why the Earth had a magnetic field in the first place. Where the poles were seemed a secondary question.

By the beginning of the 1960s, there was a growing body of scientists who had become convinced that the Earth's magnetic field had indeed periodically reversed in the past. A mechanism had also been put forward by Sir Edward Bullard for the generation of the Earth's magnetic field. Evidence suggested that the planet's outer core consists of liquid iron (see Chapter 4). Bullard proposed that flow in this liquid could cause it to behave like a natural dynamo, which in the process of generating electricity creates a magnetic field. In this model, it was perfectly possible for a change in the liquid flow to reverse the direction of the magnetic field. If the Earth's magnetic field had indeed flipped in the past, then volcanic rocks of the same age throughout the world should show the same direction of remanent magnetism. But to check whether this was indeed the case, it would be necessary to date the rocks extremely accurately.

As is often the case in science, the solution to this scientific problem required a new scientific method. This was a technique for dating rocks using the radioactive decay to argon gas of an isotope of potassium which commonly occurs in volcanic rocks (see Chapter 1). When volcanic rocks are still molten, any argon gas created by the decay of potassium can escape. But after the rock has solidified, the argon produced by radioactive decay is trapped, and starts to accumulate. Thus, the amount of argon gas in the rock today is a measure of the time since the rock was molten. However, the quantity of argon is minute and, in the early 1960s, the University of Berkeley in California was the only place in the world where sufficiently precise measurements could be made.

The scientists studying the magnetism of volcanic rocks eagerly used the potassium–argon method to date their samples. They found, as the Japanese scientist Matayuma had suggested in the 1920s, that there is a correlation between the age of a volcanic rock and its direction of remanent magnetism. This confirmed the idea that the Earth's magnetic field periodically switches – reversals appear to happen approximately once every million years or so. The stage was set for the solution to the mystery of the ocean floor.

## A RADICAL HYPOTHESIS

In 1960, Harry Hess, an academic at Princeton University in America, produced a widely circulated manuscript which speculated on the origin of the ocean floor. Hess had been fascinated by the deep ocean ever since he was a naval officer in the Pacific during the Second World War. He had taken every opportunity during his naval duties to explore the ocean floor with the newfangled echo-sounder, and he was deeply interested in the new scientific discoveries which were now emerging from research cruises. Hess realized that the rocks beneath the ocean floor must have a different composition to the rocks in the continents – indeed, they must be denser, overall, because they are at a lower level than the surface of the continents. His argument was based on the principle of 'isostasy', which assumes that the outer part of the Earth floats on the more fluid interior, just like a ship at sea. But like a heavily laden ship, the dense rocks beneath the ocean floor sink deeper into the Earth than the lighter continental crust. In fact, Hess proposed that the ocean floor beneath its thin sediment cover is made of essentially the same material as the Earth's mantle, with nearly the same high density, but altered to a mineral called serpentine where it came into contact with sea water.

Hess also had ideas about the origin of mid-ocean ridges. Scientists had found evidence for higher temperatures beneath the ridge crests than elsewhere on the sea floor – they had measured the increase in temperature within the top few metres of sediment on the mid-ocean ridges, and from this inferred that the rocks were unusually hot at depths of a few kilometres. The higher temperatures would cause the rocks to expand more, pushing up the sea floor and also lowering their density. Perhaps this was the reason why the mid-ocean ridges existed at all? They could simply be where hotter and less dense parts of the Earth's deep interior rise to the surface.

Hess went even further with his idea. One of his discoveries in the Pacific, made with the echo-sounder during the Second World War, had long nagged at him. He had found strange flat-topped and circular underwater mountains, which he called guyots. The guyots were strongly reminiscent of the many low islands in the Pacific, smoothed by the erosive power of the waves – yet the guyots were thousands of metres underwater. The heights of the guyots above the surrounding ocean floor were always about the same. Thus, as the ocean floor became deeper, so did the tops of the guyots. It occurred to Hess that, perhaps, the ocean floor is sinking all the time. Then, if the guyots were once islands, they had been carried down below sea level with the rest of the sinking ocean floor. But how did this square with his idea that the mid-ocean ridges were zones of upwelling in the mantle? Hess's solution was extremely simple. The ocean floor is part of a gigantic conveyor belt in the Earth. The mantle rises at the crest of the mid-ocean ridge, creating new sea floor, then moves sideways, away from the crest, slowly cooling, contracting and sinking all the time.

# HESS'S THEORY OF SEA FLOOR SPREADING

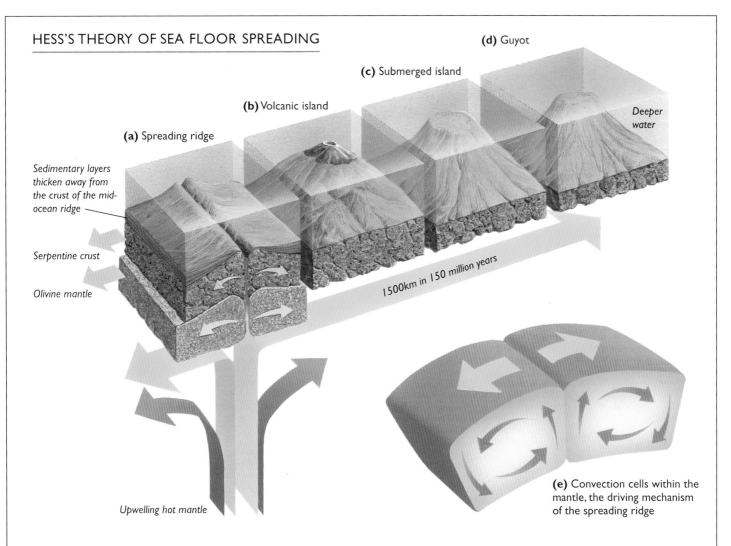

**(a)** Spreading ridge

**(b)** Volcanic island

**(c)** Submerged island

**(d)** Guyot

Deeper water

Sedimentary layers thicken away from the crust of the mid-ocean ridge

Serpentine crust

Olivine mantle

1500km in 150 million years

Upwelling hot mantle

**(e)** Convection cells within the mantle, the driving mechanism of the spreading ridge

In the late 1950s Harry Hess, an American geologist, proposed an explanation for distinctive seamounts. The mid-ocean ridge overlies a zone of deep upwelling in the Earth's mantle **(a)**. Hess thought that the upwelling hot mantle reacts with cold sea water to form ocean crust made of serpentine. The crust moves sideways, cooling and contracting at the same time, and a layer of sediment accumulates on top. In the process, the ocean floor sinks several kilometres, carrying down isolated volcanic islands **(b)**. As these sink, the powerful force of ocean waves erodes a flat platform **(c)**. Eventually submerged, these appear as seamounts or guyots **(d)**. Hess suggested that the movements in the mantle and sea floor are part of large-scale convection cells inside the Earth, which are in constant motion **(e)**.

This way, the flat-topped mountains had been carried down below sea level, finally to become underwater guyots.

The term 'sea floor spreading' was soon coined to describe Hess's ocean conveyor belt. Radical as this idea was, it explained a surprising number of phenomena. It required much of the ocean floor to be geologically young, something that was already hinted at by the results of the MidPac expedition. The sideways motion of the sea floor, away from the mid-ocean ridges, would explain why the crest of the mid-ocean ridge often looks like a gigantic rift in the Earth's crust – the ocean floor is splitting apart here. Sediment on the sea floor becomes thicker away from the mid-ocean ridges; this is simply because the sea floor is older here with more time for sediment to accumulate.

In 1962, Harry Hess was invited by the geology undergraduates at Cambridge University in England to talk about his new ideas. He gave a spell-binding performance. Like a masterful magician, Hess seemed to be conjuring up the secrets of the deep. Fred Vine, an undergraduate at the time, was deeply impressed by this talk. That same year, now as a graduate student at Cambridge, he was given the job of analysing some of the results of a recent British magnetic survey in the Indian Ocean near the Carlsberg mid-ocean ridge, named after the brewery which funded the first scientific expedition to this region. He tackled one aspect of the ocean floor that Harry Hess had not tried to explain: the mysterious pattern of magnetic anomalies.

It was well known that almost all oceanic islands are essentially volcanoes. Hess himself, with his discovery of guyots, had found the remains of many underwater volcanoes. Two underwater conical mountains called seamounts, which Vine also took to be extinct volcanoes, had been surveyed near the Carlsberg Ridge. The strength of the magnetic field over one of them was significantly greater than over the other. Vine found that he could easily explain this if he assumed that one seamount had erupted when the Earth's magnetic field was reversed, acquiring a reversed remanent magnetism when the lava cooled, while the other had erupted when the Earth's field had the same polarity as today, with a normal (i.e. present-day) remanent magnetism. The oppositely magnetized seamounts will add to or subtract from the Earth's magnetic field, producing the observed difference in strength. Vine went one step further. Using an early type of digital computer he was able to calculate the expected magnetic anomalies for this situation and show that they did indeed fit with the observations.

Vine made the logical deduction that perhaps all of the ocean floor, underneath its thin veneer of sediments, was volcanic rock which had erupted from the underlying mantle – Hess had merely suggested that the ocean floor consisted of mantle rock altered by the chemical action of sea water. Combining this modification to Hess's idea of sea floor spreading with his results from the Carlsberg Ridge and one final key observation – that the stripy pattern of magnetic anomalies in the oceans lay parallel to the mid-ocean ridge – Vine, together with his supervisor Drummond Matthews, came up with a theory of breath-taking simplicity. This was published in the scientific journal

*Nature* in 1963 in rather guarded terms, but the implications were clear. Vine and Matthews, following Hess's original idea, proposed that the ocean floor either side of the crest of the mid-ocean ridge glides apart like the belts on two gigantic conveyors (see opposite). The gap formed in this way is filled all the time by volcanic eruptions, creating new ocean floor. When the volcanic rock cools, it becomes magnetized parallel to the Earth's ambient magnetic field. This new ocean floor will also be carried away from the mid-ocean ridge as part of the process of sea floor spreading. At some later time, the Earth's magnetic field might reverse and the ocean floor being created will be magnetized in the opposite direction. This way, alternately magnetized stripes are built up on the ocean floor, which locally increase or decrease the strength of the Earth's magnetic field, producing the observed zebra-like magnetic anomalies. Vine pointed out that the magnetic anomaly found along the crest of the mid-ocean ridge always suggests that the rocks here have been magnetized in the present-day magnetic field, exactly what would be expected in his model if new ocean floor is forming today along the ridge axis. He was essentially saying that the ocean floor is behaving very much like a tape recorder, picking up the past history of reversals in the Earth's magnetic field. Vine later recalled, when interviewed about his theory: 'for me it was a fairly small leap to put…together…spreading and reversals because…I believed in spreading, or wanted to believe in spreading. If you worked very closely with the data…then I think you'd realize that almost out of desperation there was no other way of interpreting them.'

## MORE EVIDENCE IS NEEDED

One might think that the evidence was overwhelming for the ideas of Harry Hess and Fred Vine. So much seemed to have been explained. This was not how it was perceived at the time. Their ideas were so extraordinary, and even bizarre, that each bit of

# STAGES IN THE FORMATION OF AN OCEAN BASIN

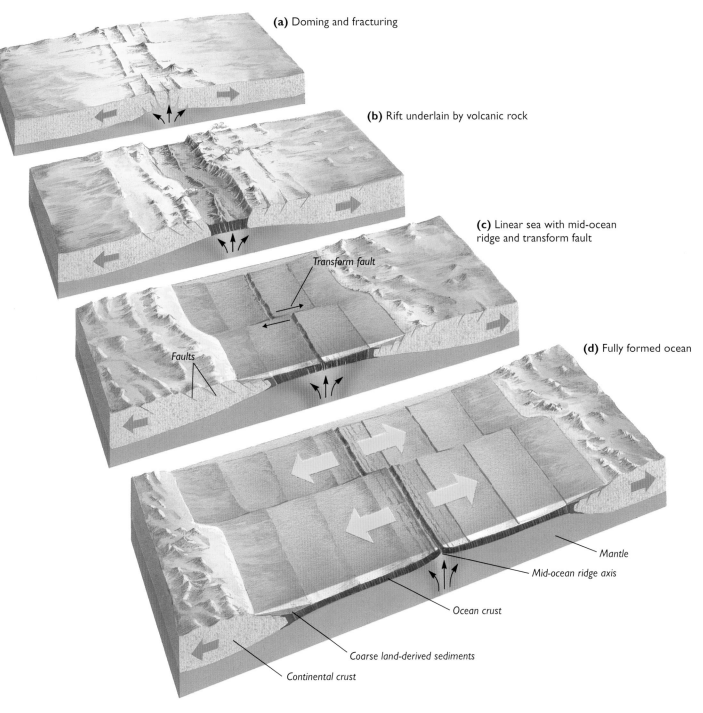

**(a)** Doming and fracturing

**(b)** Rift underlain by volcanic rock

**(c)** Linear sea with mid-ocean ridge and transform fault

*Transform fault*

**(d)** Fully formed ocean

*Faults*

*Mantle*

*Mid-ocean ridge axis*

*Ocean crust*

*Coarse land-derived sediments*

*Continental crust*

Oceans are created when a continent splits apart. In the early stages, the continent domes up and fractures, eventually creating a system of rift valleys **(a–b)**. The underlying hot mantle rises up towards the surface, triggering volcanic eruptions. As rifting continues, the crust is stretched horizontally and thinned vertically along faults. Eventually the continental crust separates and a deep linear depression is created, underlain mainly by volcanic rock. A new ocean is born when this region is flooded by water. As this widens by sea floor spreading, an original kink or bend in the rift may develop into a transform fault, which offsets the mid-ocean ridge **(c–d)**.

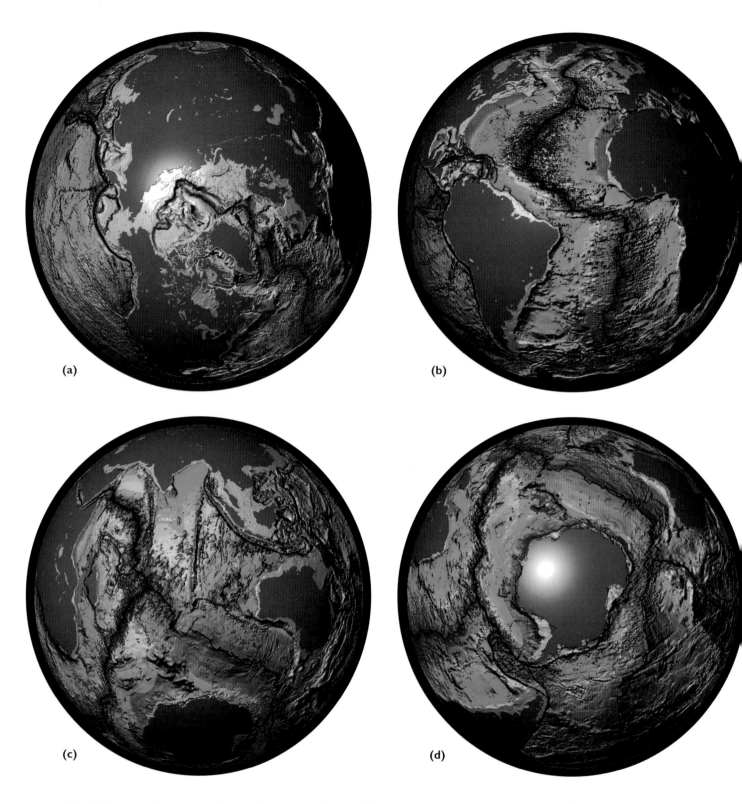

(a)

(b)

(c)

(d)

Scientists have used the pattern of magnetic anomalies in the oceans to work out the age of the ocean crust. These views of the bottom of the Arctic Ocean **(a)**, Atlantic Ocean **(b)**, Indian Ocean **(c)**, and oceans around Antarctica **(d)** show that the ocean crust is youngest (coloured red-orange) along the crest of the central mid-ocean ridge, and becomes progressively older (coloured yellow-green) on the flanks towards the ocean margin. The oldest ocean floor is approximately 180 million years old.

evidence, when scrutinized by scientists who were predisposed to be highly critical, was not as solid as Hess or Vine could have wished – the available data for the oceans were still poor. But if there was not a strong wind of change blowing through geological circles, at least there was a light breeze.

J. Tuzo Wilson, an enthusiastic geophysicist from the University of Toronto, visited Fred Vine in Cambridge in 1965, and started to examine the implications of his work. If the oceans grow by spreading at the mid-ocean ridges, producing alternately magnetized sea floor, then one should observe the same pattern of magnetic anomalies either side of the mid-ocean ridge. The first convincing proof of this was obtained the following year, when the US research vessel *Eltanin* measured a magnetic profile across the mid-ocean ridge in the Pacific Ocean, south of Easter Island, extending for over 4000 kilometres (see p. 45). The profile shows an almost perfectly symmetrical pattern of magnetic anomalies, extending either side of the mid-ocean ridge for over 2000 kilometres. If the profile is folded about the mid-ocean ridge, so that one side overlies the other side, the agreement is staggering. Also, every one of the reversals of the Earth's magnetic field, which had been so painstakingly worked out on land, could be matched with magnetic anomalies in the *Eltanin* profile.

Tuzo Wilson also made sense of the confusing pattern of fracture zones and mid-ocean ridges. He proposed that the crest of a mid-ocean ridge and the fracture zones together form a single zigzagging boundary between two pieces of oceanic crust which are moving away from each other. Thus, as the ocean floor spreads sideways at the ridge, it also slides past adjacent sections of ocean floor along the fracture zones (see p. 40). The significance of this is best seen by ripping a piece of paper in two with a zigzag tear, placing the two sides of the tear together again on a flat surface, then slowly sliding them apart so that the pieces move parallel to some segments of the tear – here fracture zones form. Ridges form along the other segments where the motion is at right angles to the tear. Wilson predicted that the movement along the fracture zones would be confined to sections which

linked ridge segments – he called this type of fault a transform fault. Here, parts of the ocean floor either side of the fracture zone would slide past each other with a predictable motion. When scientists started to accurately locate the earthquakes in the oceans – each one triggered by movement on a fault in the crust – they found that they lay along a zigzagging line, just as Wilson had predicted, following both segments of the crest of the mid-ocean ridge and the transform faults. In one fell swoop, all the puzzling features of the sea floor had been explained away.

Allan Cox, one of the pioneers of the study of reversals in the Earth's magnetic field, once remarked that 'the structure of the sea floor is as simple as a set of tree rings, and like a modern bank cheque it carries an easily decipherable magnetic signature'. Using the dated history of reversals, laboriously worked out on land, geophysicists were able in a few years to map the age of the ocean floor, which makes up nearly two-thirds of the Earth's surface, with an accuracy that, on the continents, had taken geologists a century to achieve (see opposite). The magnetic anomalies show that none of the rocks of the deep ocean floor are more than about 200 million years old. The continents are on average ten times older than the present ocean basins. Today, all along the crest of the mid-ocean ridge, new ocean floor is being created, adding about 3.5 square kilometres of new crust to the planet every year.

## DRILLING THE SEA FLOOR

Throughout the history of the exploration of the oceans, scientists have been frustrated by their inability to get their hands on the actual 'stuff' which makes up the sea floor. After all, many of these scientists had been originally trained as land geologists, spending the early part of their research careers hitting lumps of rock with a hammer and taking samples back to the laboratory. Sea geology had to be done essentially at a distance, using various remote-sensing techniques such as echo-sounding, magnetic and gravity measurements. In the mid-1960s, a special ship was adapted with a drilling rig to become the most sophisticated ocean-going research vessel in the world, capable of drilling holes in the floor of the deep ocean and bringing up long cylinders of rock called cores. The ship was called the *Glomar Challenger*, after both the company that built it (Global Marine Company) and HMS *Challenger* which undertook the first great voyage of ocean research in the 1870s. The subsequent research carried out by the *Glomar Challenger* probably ranks

as one of the single most important experiments in the history of the study of our planet.

The initial eighteen-month cruise of the *Glomar Challenger* in 1968 confirmed beyond a shadow of doubt the predictions of the sea floor spreading concept. The third leg of this voyage involved drilling nine holes in the sea floor of the South Atlantic, between South America and southern Africa, where ocean depths range between 2 and 4.5 kilometres. The holes penetrated over one hundred metres into the sea floor, reaching the volcanic rock basalt, and

## 'GLOMAR CHALLENGER' DEEP SEA DRILLING

*Numbers 1–7 refer to drill holes in sea floor*

In 1968 a specially designed ship – the *Glomar Challenger* – drilled a series of holes into the floor of the South Atlantic Ocean **(a)**. Long cylinders of rock extracted from the holes revealed the nature of the top part of the ocean crust **(b)**. Layers of sediment rest on top of basalt. Away from the axis of the mid-ocean ridge, the age of the deepest sedimentary layer becomes progressively older **(c)**. These findings confirmed the model of sea floor spreading for the creation of the ocean crust.

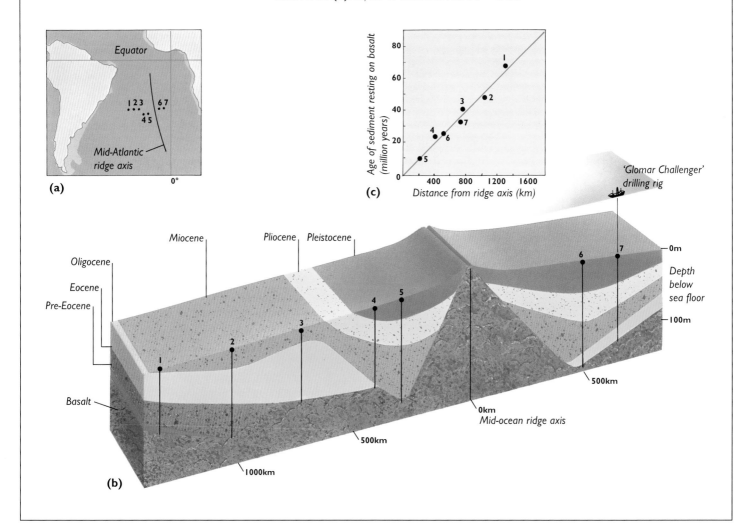

*Scientists have greatly improved their understanding of the origin of the oceans by drilling into the deep sea floor, pushing drilling technology to its limits. The JOIDES Resolution can drill in water over 6 kilometres deep, and is constantly in use throughout the world's oceans on scientific research cruises.*

long rock cores were successfully recovered. This confirmed that there was volcanic rock beneath the ocean floor. Above the volcanic rock, the cores consisted of layers of deep sea sediments full of the remains of microscopic organisms. These organisms can be dated by comparing them with organisms found in dated rock sequences on land. If the sea floor spreading model of ocean-floor creation is correct, then sediments would accumulate on the volcanic ocean crust soon after its formation. Thus the oldest organisms in the cores, found in the sediment layers immediately above the volcanic layer, should have a similar age to the underlying volcanic rock, and become progressively older away from the mid-ocean ridge. And this indeed is what was found. If these ages are compared precisely with the distance away from the ridge, it is found that they are directly proportional to this distance – exactly what would be expected if the ocean floor was spreading at a constant rate. In the South Atlantic, new ocean floor is being created at a roughly constant rate of about 4 centimetres per year in directions perpendicular to the mid-ocean ridge. As this happens, Africa and South America are moving away from each other.

The corollary of sea floor spreading in the Atlantic is that in the past Africa and South America were closer together. When the very oldest South Atlantic

Ocean floor was being created about 125 million years ago, the continental margin of Africa and South America must have been nearly touching (the margin is not the coastline but the edge of the shallow continental shelf where the deep ocean starts). No wonder the margins of these two continents look similar – Africa and South America were once one and have rifted apart (see p. 56). This was the very idea that Alfred Wegener had put forward in 1915, to be dismissed by most of the scientific community. This was the idea that Sir Edward Bullard had demonstrated so elegantly in 1965 with his computer fit of the continents. And now it had been proved, not by evidence on land, but rather by drilling in the middle of the ocean. The deep sea drill holes showed more. The sequence of sediments which overlie the volcanic crust record the history of the Atlantic Ocean, from the moment it formed as a narrow rift between two continents, rather like the great rift valley system in East Africa, to a wide ocean several kilometres deep.

By the end of the 1960s, geologists knew enough about the nature of the ocean floor to begin to recognize fragments that somehow had been pushed up above sea level. One example is the island of Iceland in the North Atlantic where the mid-ocean ridge is actually above sea level. Here, there is constant volcanic activity, earthquakes and evidence

# THE OPENING OF THE ATLANTIC

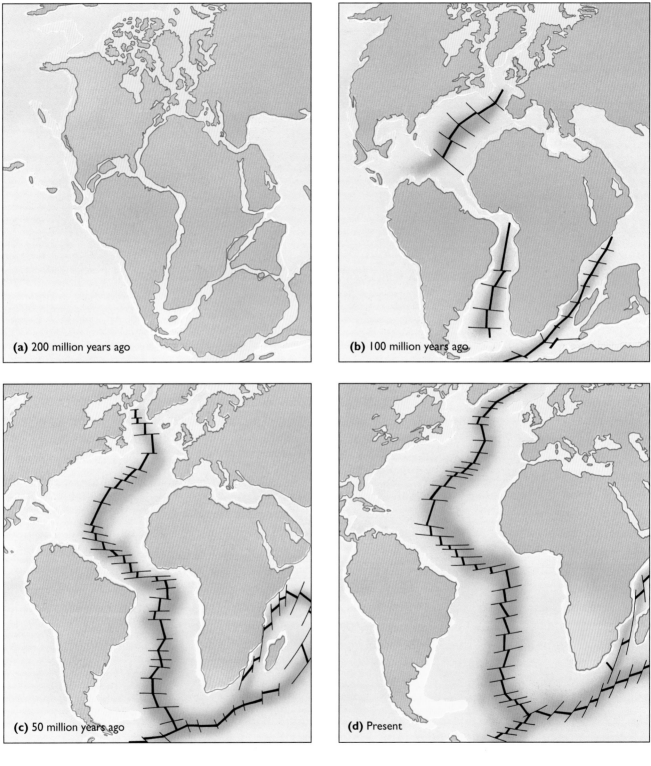

**(a)** 200 million years ago

**(b)** 100 million years ago

**(c)** 50 million years ago

**(d)** Present

About 200 million years ago the Atlantic Ocean did not exist and the continents were welded together **(a)**. The Central Atlantic Ocean was born about 150 million years ago as North America split from Africa – the gap created was filled with new ocean crust by the process of sea floor spreading at the mid-ocean ridge. Subsequently, the South Atlantic formed as South America drifted away from Africa **(b)**. Eventually, a continuous ocean extended from Greenland to the tip of Africa **(c)**, which increased in width to its present-day size **(d)**.

*The mountains in central Oman are the remains of the deep sea floor where it has been pushed up on land during large-scale crustal movements. Geologists call these rocks ophiolites; they provide many clues to the origin of the ocean crust.*

for stretching of the crust. During volcanic eruptions long fissures open up, releasing molten basalt. The mechanism of sea floor spreading is clearly evident, though the crust in Iceland is much thicker than that in the oceans. In fact, this is not a typical site of ocean crust formation and seems to be related to the upwelling of an unusually hot plume in the Earth's interior (see Chapter 4). Elsewhere in the world, fragments of basalt and deep sea sediments are sometimes found in mountain belts. How they got there is still controversial; nonetheless they provide an accessible view of what lies beneath the ocean floor, impossible to obtain any other way. One of the best examples is in the deserts of Oman, where ophiolite, a distinctive rock sequence resting on mantle rocks, is found. This can be thought of as the remains of different levels in a submarine volcano which erupted along the axis of a mid-ocean ridge.

In a typical ophiolite, the uppermost part consists of a layer of chert, a few hundred metres thick, which

deep sea silica-rich oozes have turned into rock. These sediments rest on a layer of basalt which has a characteristic pillowy shape, typical of lava erupted into water. Beneath the pillow lavas are vertical sheets of basalt, called dykes, which have cooled in gigantic fissures – clear evidence of the sea floor moving apart and molten rock welling up to fill the cracks. The dykes are merely the deeper parts of the eruptions which feed the pillow lavas. At even deeper levels, there is another layer with a similar composition to basalt, where molten rock has ponded and cooled slowly. All these layers form a crust, about 7 kilometres thick, which rests on mantle rock full of the green mineral olivine. By measuring the speed of seismic vibrations in these rocks, geologists have been able to directly compare the layers revealed by seismic probing of the sea floor with the on-land fragments. Thus, in a curious twist of scientific research, land geologists have filled in our picture of the deep sea floor.

## MODELLING THE SEA FLOOR

One important feature of the sea floor spreading model was not resolved until the mid 1980s. This was the reason why the upwelling mantle melts in the first place to form the basaltic volcanoes which build up the oceanic crust. Geophysicists have known for a very long time that the Earth's mantle is solid; it rings like a bell during earthquakes. The upwelling mantle beneath the mid-ocean ridge is a form of solid state flow. This type of flow happens very slowly and is similar to the sort of flow that lead or glass exhibits if left for long enough (see Chapter 4). The mantle is very hot but, because of the huge weight of the overlying rock, it is at very high pressures and this keeps it solid. Dan McKenzie, a geophysicist at Cambridge University, realized that as the hot mantle rises beneath the axis of the mid-ocean ridge, it steadily decreases in pressure as it gets nearer the surface. Eventually, it reaches a sufficiently low pressure that it can no longer remain solid and so it melts. The molten rock erupts in the median valley of the mid-ocean ridge to build up the ocean crust. The thickness of this crust merely reflects the amount of volcanic rock that can be generated by de-compression of mantle rock rising to the Earth's surface – in this case a thickness of about 7 kilometres. Because the deep ocean crust throughout the world has all been created by the same process, it all has the same thickness.

Sea floor spreading is not confined just to the oceanic crust. The underlying mantle is also in motion, moving horizontally away from the crest of the mid-ocean ridge. The mantle, like many other materials, becomes stronger as its temperature decreases, and below a critical temperature is much stronger than the rest of the Earth. This strong part of the Earth is called the lithosphere. It generally extends down to the depth where the mantle is at a temperature of about 1300°C. Below the lithosphere, the mantle is hotter and weaker in a region geologists call the asthenosphere (*astheno* means weak). The sea floor is kept cool (nearly zero degrees centigrade) by the overlying cold ocean water. The mantle beneath the crest of the mid-ocean ridge is very hot – at a temperature very close to the melting point of the basalt which erupts to form the oceanic crust. But eventually, as it moves away from the ridge, the mantle cools down. At the crest of the mid-ocean ridge, the lithosphere has virtually no thickness at all, and the top of the asthenosphere almost reaches the surface. But further away from the crest, as the mantle cools, the lithosphere becomes progressively thicker, reaching a maximum thickness of about 120 kilometres beneath the deep and old ocean floor.

The sea floor becomes progressively deeper away from the crest of the mid-ocean ridge. This can be explained by contraction of the lithosphere as it cools. The more the lithosphere cools, the greater is this contraction. In the 1960s scientists calculated, using the results of experiments on mantle rocks, that the depth of the ocean floor should be proportional to the square root of its age. It is testimony to the power of the idea of sea floor spreading that this mathematical relation can predict the actual shape of the sea floor over much of the world's oceans.

## JOURNEY TO THE BOTTOM OF THE SEA

In the 1970s, the desire to actually visit the mid-ocean ridge finally became irresistible. This region is truly the birthplace of most of the planet's surface. By this time, sonar systems had been developed to map the sea floor with immense accuracy. These were designed to be towed by a ship just above the sea floor and send high frequency sound pulses out sideways. Because the sound travels at a low angle to the sea floor, the image produced by picking up the echoes is rather like a photograph lit with a very low-angle light source, and even low ridges or depressions cast a pronounced shadow. Computers can process this information to produce a three-dimensional image.

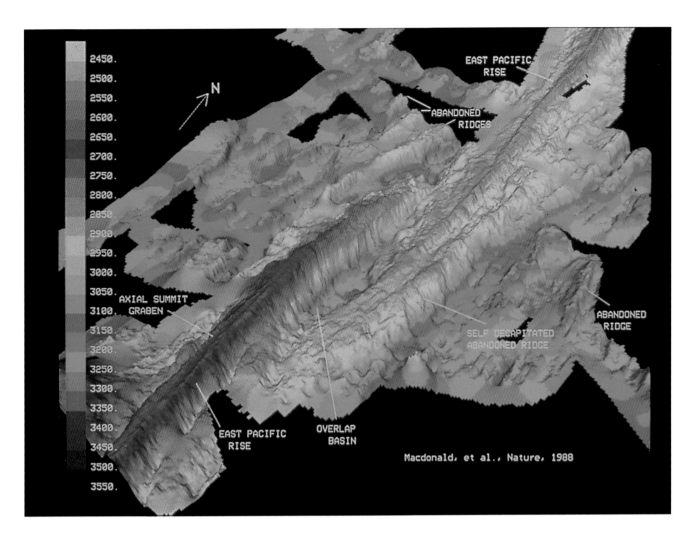

*This 3-D image of the crest of the mid-ocean ridge in the eastern Pacific (colour-coded for depth in metres below sea level) has been generated from detailed underwater sonar surveys. The ocean crust is created by volcanic activity in the axial depression (referred to as summit graben). In detail, the ridge crest is divided into a series of segments.*

The scientists who first examined the new detailed computer-generated images of the ocean floor probably felt very much like astronauts looking at satellite pictures of the surface of another planet. The time was right for a space shot, not to outer space, but to the inner space of the deep ocean. This involved diving down to depths of over 3 kilometres, where the water pressure was at least 300 atmospheres. To withstand these pressures, a very special type of submarine is needed. Today, there are only about half a dozen research submarines in the world which are capable of diving to the deep sea floor. The Russians operate the MIR submarines, and the French and Americans dive in the *Nautile* and *Alvin*.

The descent to the deep ocean floor usually lasts about two hours – the time taken to fall with a spiralling motion through more than 3 kilometres of sea water. The titanium hull resists the full force of the deep ocean, and the interior of the submarine is maintained at atmospheric pressure. The position of the submarine relative to the bottom is monitored with an echo-sounder and, when the sea floor is close, the descent can be halted by jettisoning weights to increase buoyancy. From then on, the submarine relies on its own battery-powered motor. The pilot and scientists can observe the sea floor, lit by the submarine's powerful lights, through remote video cameras and the small windows. In this way, the submarine travels through the strange landscape of the ocean bottom in a region which must surely be the remotest on Earth. During a single dive, the 'bottom time', during which the scientists can explore the sea

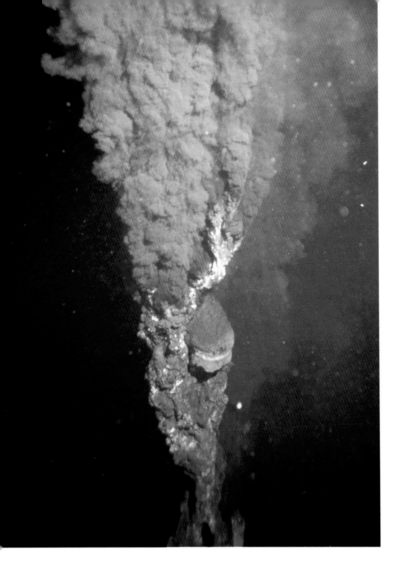

*Black smokers are an important feature of the crest of the mid-ocean ridge. These occur where superheated water, heated by hot rocks at depth, gushes out of the sea floor. When the hot water mixes with cold sea water, minerals precipitate, creating a black cloud.*

patches of water rising from cracks in the lava which were slightly warmer than the general near zero temperatures of the deep ocean. But in 1979, during a dive on the mid-ocean ridge in the Pacific Ocean, a joint French–American expedition discovered something completely unexpected. They saw a black cloud rising up from the sea floor. A thermometer, manoeuvred into the centre of the black cloud, promptly melted. This caused some alarm as the maximum temperature that the thermometer could withstand was over 300°C, close to the maximum temperature that the submarine's perspex windows could measure, and the submarine had narrowly avoided going straight through the black cloud! An incredible temperature of 350°C was eventually measured – a superheated plume of muddy water was clearly gushing out of the sea floor. This type of plume was christened a black smoker, and many more have subsequently been found along the axis of the world's mid-ocean ridges.

Black smokers are visible evidence of the flow of sea water through the oceanic crust, but this may be occurring in a much wider region up to 1500 kilometres from the ridge axis. This activity has turned out to be of fundamental importance in regulating both the cooling of the ocean floor, as it moves away from the ridge axis, and also the chemistry of the oceans themselves. Cold sea water penetrates deeply into the crust through cracks. It is then heated by hot rock at depth. The superheated water, being much more buoyant than cold sea water, rises up rapidly to escape again from the sea floor. It has been estimated that in a few tens of millions of years, the entire volume of the world's oceans is flushed through cracks in the ocean floor. As the circulating sea water is heated, so the surrounding rock is cooled, and the two react with each other. The volcanic rock is chemically altered and new minerals form which contain water in their structure. Salts, dissolved in sea water, are extracted from the oceans, but elements are also scavenged from the volcanic rock. Where the hot water reaches the sea floor, some of the scavenged elements are deposited again. Indeed, the deep sea dives have shown that fabulously rich mineral deposits exist at the bottom of the ocean. These

floor, is about four hours. When this has expired they must jettison more metal weights to allow the submarine to ascend to the support ship, reaching the sea surface two hours later.

The early submarine dives confirmed that the axis of the mid-ocean ridge is the site of intense volcanic activity. Evidence of young lava flows was everywhere. These have erupted like oozing toothpaste to form the characteristic tubular and pillow shapes of under-water lavas. On one occasion in the Pacific, the scientists detected signs that there had been a volcanic eruption just before their dive – water shimmering above the still warm lava. Occasionally, temperature probes on the submarines would detect chemical exchanges balance the influx of salts and other ions brought into the oceans continuously by rivers. Were it not for this, the oceans would have long ago become a chemical soup, probably too salty and alkaline for marine life to have any chance of survival.

## LIFE IN THE DEEP OCEAN

The discovery of black smokers coincided with another extraordinary find: clustered near the base of the hot plumes are numerous sea creatures that seem to be surviving quite happily in what would be to most organisms extremely hostile conditions. Hundreds of different species have now been recognized near the hot water vents, many of them species of clam, mussel, shrimp and tube worms new to science. The shrimps are blind, but have heat sensors. Some of the worms are several metres long. Interestingly, the shrimps, collected from round the base of the black

*The very deep ocean floor was once thought to be devoid of life. In the last twenty years, scientists have found organisms, such as these tube worms, living in the mineral-rich warm water at the base of black smokers along the crest of mid-ocean ridges.*

smokers at pressures of 300 atmospheres, survive in containers of water at atmospheric pressure in the laboratory.

When the vent faunas were first discovered, scientists were puzzled as to how they could survive. What could they be living on? Further investigation showed that along with the larger creatures, small single-celled organisms thrive either in suspension or on rock surfaces near the vents, forming the bottom of the food chain. They can survive at much higher temperatures than most organisms, and for this reason are called hyperthermophiles (ultra-high temperature lovers). They can live in water at temperatures of 100°C – the boiling point of water at atmospheric pressure – eating an unusual food. In fact, they get their energy from the hydrogen sulphide in the black smoker when it reacts with dissolved carbon dioxide and oxygen in ordinary sea water – a unique mode of feeding called chemosynthesis.

The unexpected discovery of living communities along the mid-ocean ridge has opened up a tantalizing possibility. Not only are the mid-ocean ridges the birthplace of the oceanic crust which covers most of the Earth's surface, but they may be the birthplace of life itself. Studies of the genetic material (DNA) of the hyperthermophiles suggest that they are offshoots from the lowest branches of the tree of life. Their primordial nature is also suggested by the fact that the oldest-known fossils are also the remains of single-celled organisms which thrived near hot springs about 3.5 billion years ago (see Chapter 1). We will discuss in Chapter 7 how their peculiar way of life may be the key to how life started in the first place. But for the moment we can only be amazed at what an extraordinary place the bottom of the deep sea has turned out to be. The exploration of this region has brought oceanographers, geologists and biologists together, completely changing the way they view the planet. The ocean floor, far from being ancient and dead, has proved to be one of the most dynamic places on Earth, and has a huge influence on the environment we live in.

But there were further major discoveries to be made, described in the next chapter, before scientists began to get a complete picture of this dynamism.

# CHAPTER 3

# RING OF FIRE

································································

The Pacific Ocean is rimmed by a chain of active volcanoes, arranged in a series of graceful arcs and extending 30,000 kilometres from New Zealand through Fiji, New Guinea, the Philippines, Japan, the Aleutian Islands, and down the west coast of the Americas to Patagonia. This necklace of volcanoes, continually rocked by earthquakes, has been christened the 'Ring of Fire'. Scientists exploring the link between the Pacific Ocean and the earthquakes and volcanoes which surround it have formulated a remarkable theory, plate tectonics, which explains not only how the outer part of the Earth works, but how the continents themselves, and the mineral wealth they contain, were first formed and continue to grow.

*A volcanic eruption in the Ring of Fire. A column of hot gas and ash rises from the summit of Mount St Helens in North America during the second eruption of 1980, a month after the lethal sideways explosion which devastated the surrounding region.*

## A MOUNTAIN OF RICHES

The first Europeans to set eyes on the Pacific Ocean were probably Spanish conquistadors, who fought their way down the west coast of South America looking for land, status and riches – things they could never have in the old country of Spain. By 1535, the conquistadors had reached the high Andean lands of what is today Bolivia, which they called Upper Peru. Here, in a rugged region of snow-capped mountains reaching over 6000 metres above sea level, where the air has half the oxygen as that at sea level and the night-time temperatures can fall to 30 degrees below zero, the conquistadors heard rumours of fabulous mineral wealth – thick veins of silver metal that were visible in the bare rocky ground. One mountain with a distinctive conical shape stood out from all the rest in a region the locals called Potosi. It harboured silver deposits the like of which the conquistadors had never seen. They christened it 'Cerro Rico' – the mountain of riches.

At the foot of the mountain the Spanish founded the city of Potosi, which over the next 300 years they endowed with ornate churches and palaces. But the heart of Potosi was the Casa de la Moneda, a massive fortified stone building where silver, extracted from the Cerro Rico, was turned into coins. The coins minted here formed a river of silver which flowed back to Spain: enough money got back to finance an empire. It has been estimated that during the 300 years of Spanish colonial rule in South America, 2000 million ounces of silver were mined from the Cerro Rico. This is more than all the silver mined elsewhere in the world during the same period, and is a substantial fraction of all the silver in human hands.

*Local people celebrate the founding of their town of Potosi in Bolivia. The wealth of the town comes from the fabulously rich silver mines in the Cerro Rico, a distinctive conical hill which is clearly visible in the background. It is estimated that 2000 million ounces of silver have been extracted from this mountain.*

# THE RING OF FIRE

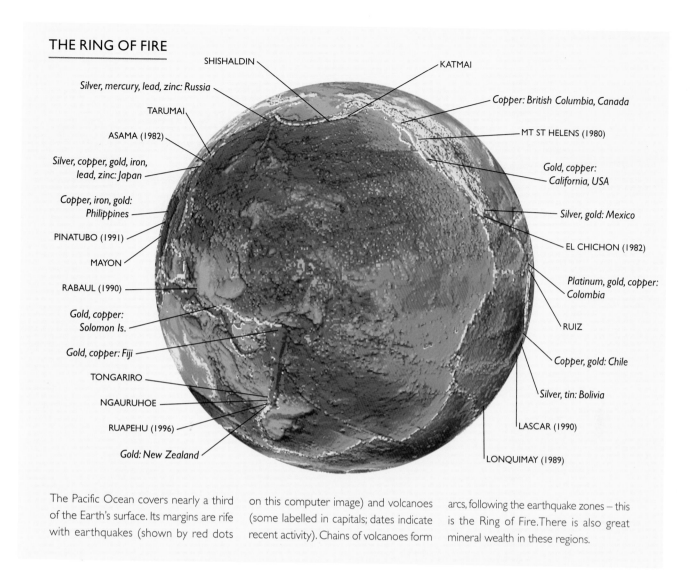

SHISHALDIN

KATMAI

Silver, mercury, lead, zinc: Russia

Copper: British Columbia, Canada

TARUMAI

MT ST HELENS (1980)

ASAMA (1982)

Silver, copper, gold, iron, lead, zinc: Japan

Gold, copper: California, USA

Copper, iron, gold: Philippines

Silver, gold: Mexico

PINATUBO (1991)

EL CHICHON (1982)

MAYON

Platinum, gold, copper: Colombia

RABAUL (1990)

Gold, copper: Solomon Is.

RUIZ

Gold, copper: Fiji

Copper, gold: Chile

TONGARIRO

Silver, tin: Bolivia

NGAURUHOE

LASCAR (1990)

RUAPEHU (1996)

Gold: New Zealand

LONQUIMAY (1989)

The Pacific Ocean covers nearly a third of the Earth's surface. Its margins are rife with earthquakes (shown by red dots on this computer image) and volcanoes (some labelled in capitals; dates indicate recent activity). Chains of volcanoes form arcs, following the earthquake zones – this is the Ring of Fire. There is also great mineral wealth in these regions.

The silver from the Cerro Rico is only part of the fantastic wealth that the high Andean countries of Spanish South America have produced. The huge tin deposits of Bolivia flooded the world market in the early part of this century, bankrupting the Cornish tin mining industry and providing much of the world's tin. These regions once again dominate the world mineral markets, this time with copper. The big bodies of copper at Chuquicamata and Las Encondidas in northern Chile form a substantial part of the world's copper reserves. And among the volcanoes which encircle the Pacific Ocean, making up the Ring of Fire – in New Zealand, New Guinea, Alaska, North and South America – recent discoveries of gold deposits have started another big gold rush. Today, one of the richest gold mines in the world is being developed in the crater of the small volcanic island of Lihir off the coast of New Guinea.

But for those who live in these places, the mineral wealth comes at a price. In all the lands on the margin of the Pacific Ocean, the ground is rocked at regular intervals by violent earth tremors. In the Bolivian city of La Paz, the Basilica San Francisco, first built in 1550, was badly damaged by an earthquake in the 1700s and had to be almost completely rebuilt. Much further south, in Chile, the city of Concepción was destroyed by an earthquake in 1848 when Charles Darwin was visiting the region on his voyage in the *Beagle*. In 1960 a similar stretch of Chile, nearly 1000 kilometres long, was badly damaged by one of the largest earthquakes ever recorded in human history. And in 1970 an earthquake in Peru triggered a landslide which killed an estimated 50,000 people in a high Andean valley.

Early this century, the last remnants of the Jesuit mission in South America, which at its height in the 1700s had almost created an independent state in the

Chaco region of the Bolivian lowlands, set up an observatory to monitor earthquakes. This observatory, called the Observatorio San Calixto, housed in the palace of a Spanish colonial family, still exists and is run by a Jesuit priest. But the scientists and priests in the Observatorio, like the rest of the population of South America, can only wait for the next earthquake.

If the Jesuits saw a link between earthquakes and the fantastic mineral wealth of the Spanish Americas, it was probably as a warning from God not to covet worldly possessions. Yet today, geologists have discovered that there is a profound connection, a discovery which is part of a revolution in our understanding of the Earth. The seeds of this revolution lay in the detailed examination of the aftermath of a great earthquake in the Ring of Fire, not in South America but much further north.

## ANATOMY OF A GREAT EARTHQUAKE

Several thousand kilometres north of Bolivia, on Good Friday, 27 March 1964, an unusually large earthquake measuring 8.6 on the Richter scale struck the Pacific coast of Alaska around Anchorage. The shaking triggered numerous landslides. High-rise buildings and bridges collapsed. Huge water waves, up to 50 metres high, engulfed coastal facilities, wiping out the port installations of Valdez and Seward in Prince William Sound and flooding large tracts of land. The force of some of these water waves was quite staggering. In Seward, a 79-tonne diesel railway engine was thrown nearly 40 metres from its tracks.

### THE 1964 ALASKAN EARTHQUAKE

In 1964 large tracts of coastal Alaska moved several metres either up or down during a great earthquake which had an epicentre southeast of Anchorage **(a)**. There was a systematic pattern to these vertical movements, with a seaward zone of uplift and an inland region of subsidence **(b)**. This can be explained by slip during the earthquake on a gigantic fault which underlies the region and only reaches the surface offshore along the axis of the Aleutian trench **(c)**. The earthquake was triggered by the movement of the Pacific oceanic crust as it slid beneath the Alaskan continental crust.

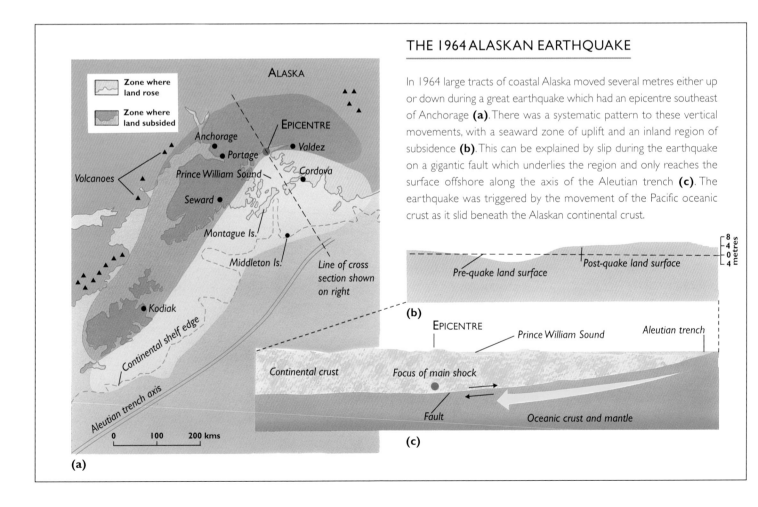

*Portage was abandoned after the 1964 Alaskan earthquake – nearly 2 metres of subsidence had made the town vulnerable to flooding during high tides. Since 1964, the region has silted up and the wooden houses, shown here in 1996, are partly buried.*

Flotsam and jetsam were flung high up on the steep hill sides.

George Plafker, who worked for the United States Geological Survey in Alaska, was attending a geological conference in Seattle when the news broke. The earthquake was in George Plafker's field area – he had been studying the geology of this part of Alaska for years. Plafker flew up to Anchorage the next day with two colleagues. The main airport had been badly damaged in the earthquake and they were diverted to a nearby military base.

Unlike the army of engineers which swarmed over the disaster area, George Plafker's scientific interest did not lie in the effects of the shaking – as a geologist, he wanted to know the cause of the earthquake. He realized that the way to do this was to work out how the earthquake had changed the landscape. Over the next few weeks, using military helicopters and float planes, Plafker explored the Anchorage and Prince William Sound regions of Alaska, interviewing eyewitnesses. He heard strange reports of land level changes. Some fishermen

thought that the land had gone up. Others spoke of flooding and subsidence. These changes were not confined to a small region, but covered an area the size of France. Clearly, something much more dramatic than just shaking had happened during the Alaskan earthquake.

The road south out of Anchorage towards Seward follows one side of Turnagain Arm, a large inlet so called because Captain Cook on his voyage of exploration to this region in 1777 found no through passage. About an hour's drive from Anchorage, you will come across the remains of the township of Portage next to the railway line. In 1964, this was a small rail and road stop, constructed out of weatherboard and reused railway rolling stock. Dick Redmond owned the local bar and shop, called Diamond Jim's. The land where he had built his bar had been carefully chosen to be well above the highest tides. But the night after the earthquake, when the tide came in, water started to ooze into the house through the crack at the bottom of the front door, flooding the floor of the bar with a few centimetres of muddy

water. The following day, the tide came in even higher, eventually reaching the level of the ground floor windows. It soon became clear that this was not a freak event. After the earthquake the township of Portage was always flooded in about a metre of water at the highest tides, and it soon turned into a ghost town. The only plausible explanation was that Portage had subsided by as much as a couple of metres. This subsidence was not confined just to Portage – large tracts of dead pine forests, killed by an influx of salt water, suggested that land levels in an extensive region south of Anchorage had also dropped (see p.66).

About 100 kilometres southeast of Portage, in Prince William Sound, the fishing harbour of Cordova suffered a different fate. Here, a small fleet of fishing boats was based, well positioned to take advantage of the annual migration of salmon. During the earthquake, buildings, water mains and electricity lines were disrupted. Paul Abbott owned a large fish cannery in Cordova. In the days following the earthquake, he recalls that there was a tremendous amount of confusion. Large masses of water were sloshing around Prince William Sound, and the fishermen of Cordova were unable to leave the harbour safely. One man, who was out on the tidal flats during the earthquake, reported how fountains of muddy water squirted up in little geysers. But a week after the earthquake, when the tidal rhythm returned to normal, the fisherman of Cordova discovered a worrying change. The high tide no longer flooded the harbour, and their moored boats were left high and dry.

It seemed that the earthquake had raised the town of Cordova out of the sea. Plafker found a simple method for measuring this uplift. Barnacles live along

*Left: Drowned pine forests near the township of Portage in Alaska. This region subsided over 2 metres and was flooded by sea water during the 1964 Alaskan earthquake.*

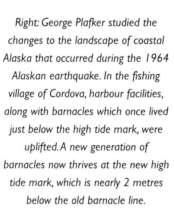

*Right: George Plafker studied the changes to the landscape of coastal Alaska that occurred during the 1964 Alaskan earthquake. In the fishing village of Cordova, harbour facilities, along with barnacles which once lived just below the high tide mark, were uplifted. A new generation of barnacles now thrives at the new high tide mark, which is nearly 2 metres below the old barnacle line.*

the rocky coastline, colonizing only rocks which are in the intertidal zone and are underwater at high tide. However, during the earthquake, these creatures had been pushed up well above the high tide levels, forming a bleached band of dead barnacles. Simply by measuring the height difference between the top of the dead barnacle zone and the new high tide mark, George Plafker could determine the coastal uplift.

Since 1964 new colonies of barnacles have grown up. In some places, at low tide, it is still possible to see two distinct bands of barnacles, with the lower-living barnacles lying several metres below the old pre-1964 barnacle zone. This is particularly clear on the forest of wooden piles which support canneries in Cordova. Here, a bleached band of dead barnacles still cling to the piles about 2 metres above the living barnacle level. The uplift was not unique to the town of

Cordova, but occurred in an extensive region, adjacent to the region of subsidence Plafker had discovered further north. In some regions, the shorelines had been raised by nearly 10 metres.

George Plafker plotted his measured changes in land level on a map. He found that he could draw smooth lines on this map which linked places that had moved vertically by the same amount. These lines were like contours on a topographic map, defining the shape of regions which had gone up or down, in an area 400 kilometres wide and 800 kilometres long between Prince William Sound, Kodiak Island and Anchorage. We can visualize this by imagining how these movements would have changed the shape of a perfectly flat surface. In the north, this surface would have sagged downwards, forming an elongated depression. To the south, a similarly shaped region would have bulged up into a dome (see p.66).

## A CONTROVERSIAL IDEA

Geologists agreed that the best explanation for the main shock was that it was a consequence of two pieces of the Earth's crust sliding suddenly past each other along a gigantic fracture or fault. Where this fault reached the surface there would be a fault line, and either side of this line the land could be expected to subside or rise. Indeed, Plafker had found just this sort of vertical movement across sharp fault breaks on Montague Island, in the middle of Prince William Sound, which had ruptured during the earthquake. Parts of Montague Island had been pushed up about 10 metres along a fault line which now nearly cut the island in two, forming a long ridge. The nature of the movement on Montague Island suggested that the floor of the Pacific was being brought closer to Alaska. But the fault break extended only a few kilometres. The local movement here could not possibly explain the regional pattern of uplift and subsidence during the earthquake. This required movement on a gigantic fault line, extending for hundreds of kilometres.

Despite all his efforts, Plafker could find no sign of a gigantic fault line on land. In an attempt to explain this, he proposed that the fault line was offshore. In other words, the fault reached the surface underwater, possibly where oceanographers had detected a deep trench in the ocean floor. Plafker went further with this idea: he speculated that the fault plane underlay almost all of coastal Alaska, sloping upwards very gently towards the south (see p. 66). During the earthquake, coastal Alaska had suddenly moved southwards and upwards over the Pacific Ocean. This seemed to geologists at the time to be an extremely radical explanation for the earthquake.

Plafker's model explained the earthquake uplift. But what about the tracts of land which had subsided? Even more puzzling, when this land was resurveyed, it was found that survey marks had shifted horizontally, moving apart by some metres. Thus, the subsiding region had actually expanded in a horizontal direction, contracting in a vertical

## ELASTIC REBOUND DURING A GREAT EARTHQUAKE

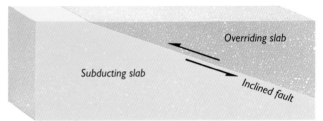

**(a)** Immediately after last earthquake

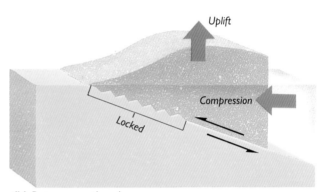

**(b)** Between earthquakes

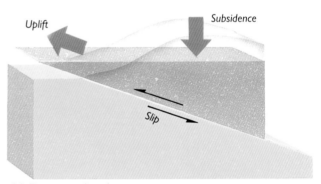

**(c)** During earthquake

Around the edge of the Pacific there is both uplift and subsidence during great earthquakes. This can be explained by movement on giant inclined faults which separate two slabs **(a)**. Between earthquakes, the overriding slab is squeezed like a spring and bulges upwards **(b)**. However, the shallow part of the fault is locked. Eventually, the locked part breaks and there is an earthquake. The squeezed crust slides suddenly upwards along the fault, expanding back to its original shape **(c)**. This pushes up the land in the toe of the overriding slab, but behind this the expanding crust subsides.

*During the 1964 Alaskan earthquake large tracts of coastal Alaska were uplifted several metres, creating a wide coastal platform. Shrubs and small trees have now colonized the new land. The shoreline before the earthquake was at the foot of the pine-tree-covered cliffs.*

direction and subsiding. At first sight there seems to be a contradiction in George Plafker's model: the earthquake was caused by the Alaskan crust moving up and over the crust of the Pacific, so that the two regions were getting closer together; but a wide region in the Alaskan crust had expanded and subsided, moving points further apart. And yet it turned out that George Plafker was right. To see why, it is necessary to appreciate the work of another geologist, Harry Reid, who had studied the great earthquake of 1906 in San Francisco.

Harry Reid suggested that the Earth's crust behaves rather like a spring. This may seem rather surprising as one may not associate rocks with springiness. The springiness is tiny and after only a very small amount of squeezing or stretching, called the elastic limit, it will break (see opposite). This behaviour is characteristic of brittle materials – in fact, the crust is brittle only where it is colder than about 350°C. At higher temperatures it behaves more like toffee and flows. There can be earthquakes only where the rocks are brittle. If one had a strip of rubber 10 kilometres long, with the same strength as the Earth's crust, then one

would be able to stretch it only between 10 centimetres and 1 metre before it snapped.

Reid saw earthquakes as part of a cycle in a sort of jerky pattern of motion. Between earthquakes, over periods of years to hundreds of years, the Earth's crust is elastically (i.e. like a spring) squeezed or stretched, until it eventually breaks. At breaking point there is an earthquake, and the crust springs back, either side of the break, to its shape just after the previous earthquake. And then the whole cycle begins again. The more the crust has been squeezed or stretched before an earthquake, the bigger the snap back will be and hence the greater the magnitude of the earthquake. This explains why very large earthquakes occur relatively infrequently, because it takes several hundred years before the crust is distorted enough to generate such a large earthquake when it breaks. What Reid did not know was the origin of the forces which were driving this cycle of elastic distortion and rupture.

Reid's earthquake cycle provided a natural explanation for George Plafker's observations. Plafker's idea was that before 1964 the whole of coastal Alaska, in a region 200 kilometres wide, had

*During the 1964 Alaskan earthquake the small island of Middleton, perched on the edge of the Pacific Ocean, was uplifted nearly 5 metres. A wrecked Second World War liberty ship, which prior to the earthquake could not be reached even at the lowest tides, was raised entirely out of the sea.*

been squeezed horizontally together like a spring and pushed up in the process. The squeezing was in a roughly northwest–southeast direction, driven by the relative motion between the Alaskan and Pacific crust. Eventually, the crust was so compressed that it broke along a gently inclined fracture. Geologists call this type of fracture a thrust fault, but Plafker called it a megathrust, because it was the largest thrust that anybody had described. Coastal Alaska expanded back to its pre-squeezing shape, sinking at the same time so that forests and small townships, like Portage, were drowned. However, the leading edge of the expanding zone was pushed slightly upwards about 20 metres over the crust in the Pacific Ocean. This region actually rose out of the sea, stranding the harbour of Cordova and parts of the rocky shoreline. Small fractures locally broke the overriding Alaskan crust, creating the fault breaks that George Plafker had seen on Montague Island.

It was clear that movement on George Plafker's megathrust was not a one-off event. Evidence for earlier events on a similar scale came from Middleton Island, a small piece of Alaska in the Prince William Sound area, perched on the edge of the deep ocean trench. The coastline of this island has a peculiar shape – it is almost like a staircase, consisting of a flight of terraces rising towards the centre of the island. The highest terrace, which runs along the middle of the island, is used as a landing strip for aircraft servicing an early-warning radar station, poised to pick up the first sign of a Russian attack.

During the 1964 earthquake, an underwater rocky platform, planed off by the rough seas around Middleton Island, was lifted out of the water to form a brand-new coastal terrace. It brought up with it the wreck of an old Second World War liberty ship which had been submerged for years in water at even the lowest tides.

George Plafker guessed that each of the higher terraces recorded earlier earthquakes. In this case, driftwood caught up on the old stranded beach terraces could be used to date the previous earthquakes. Using the radioactive carbon method of dating, he found that there was a systematic age difference between the driftwood on each terrace. This suggested that the 1964 event was not a random event, but was part of a recurring pattern which repeated itself roughly every 800 years. There was very little likelihood of another large earthquake in the same region in the immediate future.

## SUBDUCTION ZONES

Oceanographers had long known about the deep trenches in the ocean floor around most of the margin of the Pacific Ocean. It was becoming clear that the trenches are very peculiar features of the Earth's surface. They always lie a hundred or so kilometres offshore and parallel to the great arcs of volcanoes. They are also extremely deep, usually reaching 6 to 7 kilometres below sea level. Near the Mariana Islands, north of New Guinea, there is a trench which is 11 kilometres deep.

In the early 1950s, a seismologist called Hugo Benioff had found another strange feature of the trenches. He had discovered that the whole Pacific Rim region was rife with small earth tremors. Sometimes these were big enough for him to locate the source of the tremor. He found that they were not randomly distributed. In fact, they seemed to originate in a 'slab-like' zone of earthquakes, perhaps several tens of kilometres thick. This slab extended from the ocean trench as a gently inclined zone, before plunging more steeply into the Earth's interior beneath the volcanic arc, down to a depth of about 700 kilometres. These zones were soon christened Benioff zones by geologists. Benioff had proposed that they occurred where the ocean crust slid into the Earth's interior, but his interpretation had generally been ignored. In the 1950s, it made no sense to have such zones encircling the Pacific Ocean.

George Plafker realized that his proposed megathrust beneath coastal Alaska coincided with the shallow part of a Benioff zone. Soon, Plafker and fellow geologists had identified the signature of other large earthquakes, like the 1964 event, all around the Pacific Rim, in New Zealand, Japan, western North America and South America – indeed, wherever there

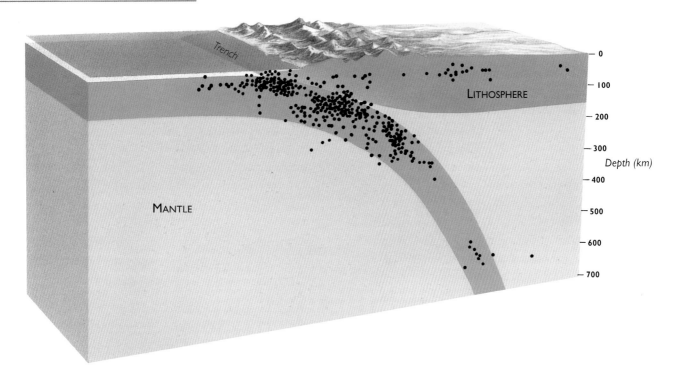

## BENIOFF EARTHQUAKE ZONE

*Trench*

LITHOSPHERE

MANTLE

— 0
— 100
— 200
— 300
*Depth (km)*
— 400
— 500
— 600
— 700

The margins of the Pacific Ocean are rife with earthquakes. Many of them (shown here by black dots) originate in 'slab-like' zones called Benioff zones after the seismologist Hugo Benioff. The Benioff zones form where the ocean lithosphere – the top 100 kilometres or so of the Earth – bends down at the ocean trench and sinks into the Earth's mantle, sliding underneath the adjacent lithosphere. As this happens, parts of the sinking lithosphere break up along faults – movement on these faults triggers the earthquakes.

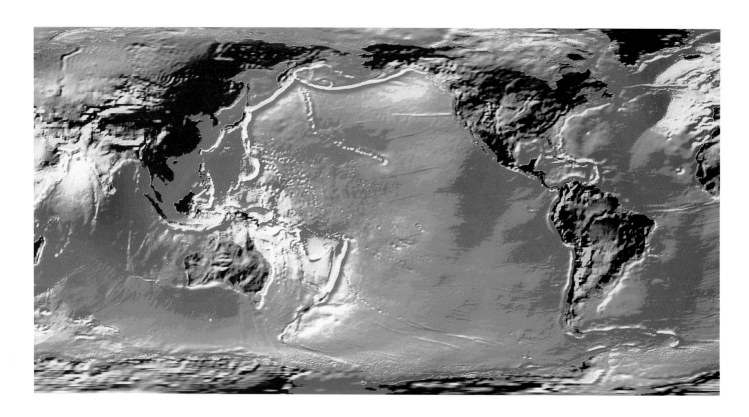

*Precise measurements of sea level – colour-coded from pale green (highest) to dark blue (lowest) – reveal the shape of the ocean floor. This illuminated relief map clearly shows deep trenches in the ocean floor, which form arcs around the margin of the Pacific Ocean where the ocean floor bends down and sinks back into the Earth's interior at subduction zones.*

was a Benioff zone. We have described the 1964 event in terms of the Alaskan continental crust sliding over the Pacific Ocean crust. But as all motion is relative, this can equally well be viewed in terms of the Pacific Ocean crust sliding or sinking beneath the Pacific margin. If the Benioff zones were part of this process, then they suggested that together with the oceanic crust, a substantial thickness of the underlying mantle must also be going down – in fact, the whole lithosphere. By 1967, it was clear that this was indeed happening. The Pacific Ocean lithosphere was sliding underneath almost all the Pacific margin. It was no longer necessary to wait for earthquakes of the magnitude of the 1964 Alaskan event to see this. Improved techniques of monitoring earthquakes, using sensitive seismometers positioned all over the world, could now be used to determine how the crust moved during the many much smaller earthquakes which were occurring on an almost yearly basis.

The origin of the deep ocean trenches now became clear. These form where the ocean crust starts to bend down into the Earth's interior, pulling down the sea floor with it. A tongue of ocean lithosphere is literally falling into the Earth's interior. Where it curves round and starts its descent, the rocks are very slightly pulled apart, giving rise to many small earthquakes. At depths of a few hundred kilometres inside the Earth, the falling lithosphere starts to feel the resistance of stiffer parts of the mantle; the lithosphere is very slightly compressed, triggering more earthquakes. Together, these earthquakes form a Benioff zone, faithfully tracking the descent of the ocean floor.

*Right: Chains of volcanoes lie above subduction zones where the ocean floor sinks back into the Earth's interior. When viewed from space, a line of these volcanoes in Java stands out clearly, stretching into the distance.*

# Where oceans die

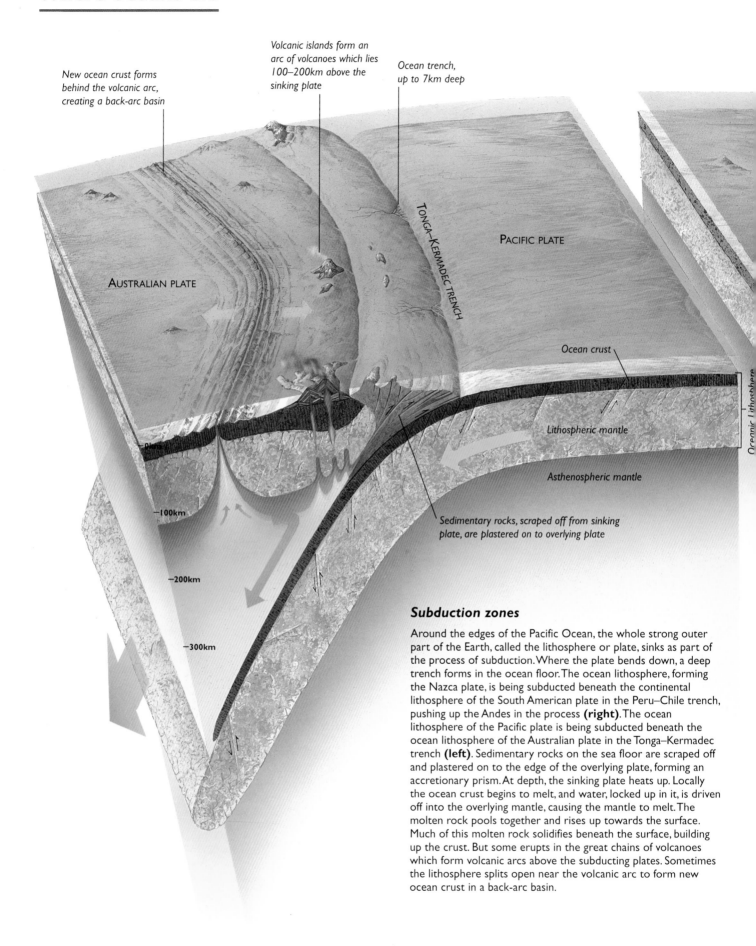

New ocean crust forms behind the volcanic arc, creating a back-arc basin

Volcanic islands form an arc of volcanoes which lies 100–200km above the sinking plate

Ocean trench, up to 7km deep

TONGA–KERMADEC TRENCH

PACIFIC PLATE

AUSTRALIAN PLATE

Ocean crust

Lithospheric mantle

Oceanic lithosphere

Asthenospheric mantle

Sedimentary rocks, scraped off from sinking plate, are plastered on to overlying plate

0km

−100km

−200km

−300km

## Subduction zones

Around the edges of the Pacific Ocean, the whole strong outer part of the Earth, called the lithosphere or plate, sinks as part of the process of subduction. Where the plate bends down, a deep trench forms in the ocean floor. The ocean lithosphere, forming the Nazca plate, is being subducted beneath the continental lithosphere of the South American plate in the Peru–Chile trench, pushing up the Andes in the process **(right)**. The ocean lithosphere of the Pacific plate is being subducted beneath the ocean lithosphere of the Australian plate in the Tonga–Kermadec trench **(left)**. Sedimentary rocks on the sea floor are scraped off and plastered on to the edge of the overlying plate, forming an accretionary prism. At depth, the sinking plate heats up. Locally the ocean crust begins to melt, and water, locked up in it, is driven off into the overlying mantle, causing the mantle to melt. The molten rock pools together and rises up towards the surface. Much of this molten rock solidifies beneath the surface, building up the crust. But some erupts in the great chains of volcanoes which form volcanic arcs above the subducting plates. Sometimes the lithosphere splits open near the volcanic arc to form new ocean crust in a back-arc basin.

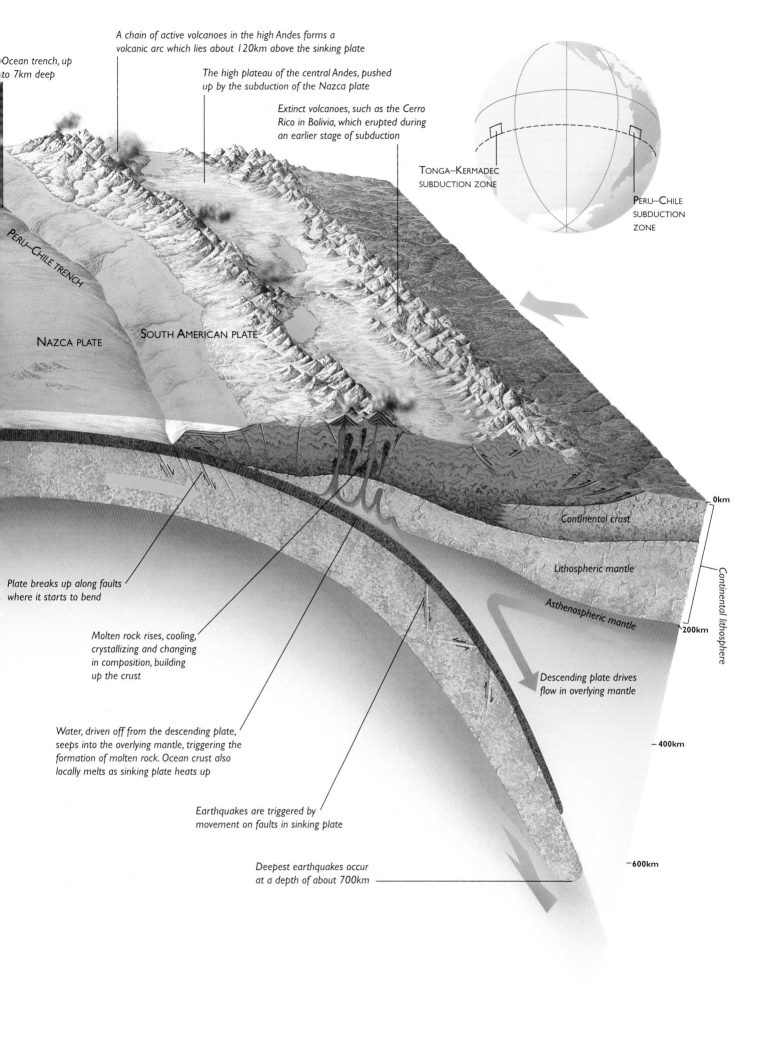

A chain of active volcanoes in the high Andes forms a volcanic arc which lies about 120km above the sinking plate

Ocean trench, up to 7km deep

The high plateau of the central Andes, pushed up by the subduction of the Nazca plate

Extinct volcanoes, such as the Cerro Rico in Bolivia, which erupted during an earlier stage of subduction

Tonga–Kermadec subduction zone

Peru–Chile subduction zone

Peru–Chile trench

Nazca plate

South American plate

Continental crust

Lithospheric mantle

Asthenospheric mantle

Continental lithosphere

—0km

—200km

Plate breaks up along faults where it starts to bend

Molten rock rises, cooling, crystallizing and changing in composition, building up the crust

Descending plate drives flow in overlying mantle

—400km

Water, driven off from the descending plate, seeps into the overlying mantle, triggering the formation of molten rock. Ocean crust also locally melts as sinking plate heats up

Earthquakes are triggered by movement on faults in sinking plate

Deepest earthquakes occur at a depth of about 700km

—600km

Geologists describe the process by which the ocean floor sinks back into the Earth's interior as subduction. The place where it happens is a subduction zone. And virtually wherever there is a subduction zone, there is a chain of active volcanoes above the sinking ocean floor. The ocean floor is not always subducted beneath the margins of the continents. Sometimes, ocean floor sinks beneath adjacent ocean. In this case, the associated volcanoes create a series of islands, arranged in an arc. The chains of volcanoes which make up many of the volcanic islands on the western margin of the Pacific Ocean are examples of these island arcs (see pp. 76–7).

## THE THEORY OF PLATE TECTONICS

At the time that George Plafker was putting forward an explanation for the 1964 Alaskan earthquake, many geologists were trying to come to grips with another set of ideas. The common-sense notion of continents which were fixed in position for all time had recently been shaken by the new discoveries in the oceans. By now, the idea of sea floor spreading had been largely worked out (see Chapter 2). Seismologists were able to show that earthquakes in the ocean crust almost exclusively occurred along either the axis of the mid-ocean ridges or along sections of the transform faults which link different segments of the mid-ocean ridge system. Large parts of the ocean floor were seismically quiet and seemed to be behaving as rigid portions of the Earth. In essence, the Earth's oceans appeared to be divided up into a series of gigantic plates, consisting of the oceanic crust and the top of the underlying mantle, together called the lithosphere. These were moving relative to each other at the axis of the mid-ocean ridges and along transform faults.

Sea floor spreading creates new ocean floor and so, on its own, must lead to an increase in the Earth's surface area. The discovery of subduction zones provided the final piece of what was turning out to be an elaborate jigsaw puzzle. If the lithosphere beneath the oceans is sliding beneath the Pacific margin, sinking back into the Earth's interior, then here was a mechanism of surface area destruction which could balance the creation of new ocean crust at the mid-ocean ridges. The Earth need neither be expanding nor contracting, but could be maintaining a constant surface area. And thus in 1967, the theory of global plate tectonics was born.

The theory assumes that the Earth has a constant surface area which today is divided up into a number of major rigid plates, and some minor ones. The plates form the strong outer shell of the Earth – the litho-

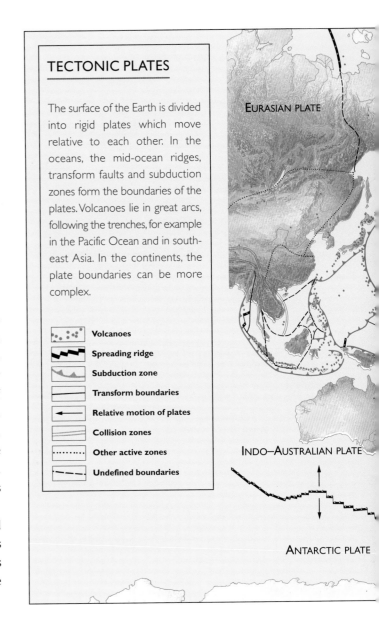

## TECTONIC PLATES

The surface of the Earth is divided into rigid plates which move relative to each other. In the oceans, the mid-ocean ridges, transform faults and subduction zones form the boundaries of the plates. Volcanoes lie in great arcs, following the trenches, for example in the Pacific Ocean and in south-east Asia. In the continents, the plate boundaries can be more complex.

- Volcanoes
- Spreading ridge
- Subduction zone
- Transform boundaries
- Relative motion of plates
- Collision zones
- Other active zones
- Undefined boundaries

EURASIAN PLATE

INDO–AUSTRALIAN PLATE

ANTARCTIC PLATE

sphere, which is on average about 100 kilometres thick. The crust makes up the top part of the lithosphere. Some plates contain only oceanic crust, others contain both continental and oceanic crust. The plates are moving relative to each other over the surface of the Earth; sometimes moving apart at the mid-ocean ridges or zones of rifting; sometimes moving past each other along transform faults; and sometimes moving together as is the case around the Pacific margin. If one thinks about it, this is such an extraordinary idea that it needs to be really tested before it is believed. The plates actually form curved portions of the Earth's surface. If they are indeed

perfectly rigid, then the surface of the Earth should move in a predictable way.

To appreciate this, imagine a ship steaming around the world along the equator. It is, in fact, moving in a circle. An observer on land might describe its track as a rotation about an axis which passes through the north and south pole. It turns out that the relative motion between any two rigid objects on the surface of a sphere, be they ships or tectonic plates, can be described in a similar way, though the axis of rotation does not necessarily pass through the north or south poles. This axis of rotation is called the Euler pole after a Swiss mathematician who investigated motion

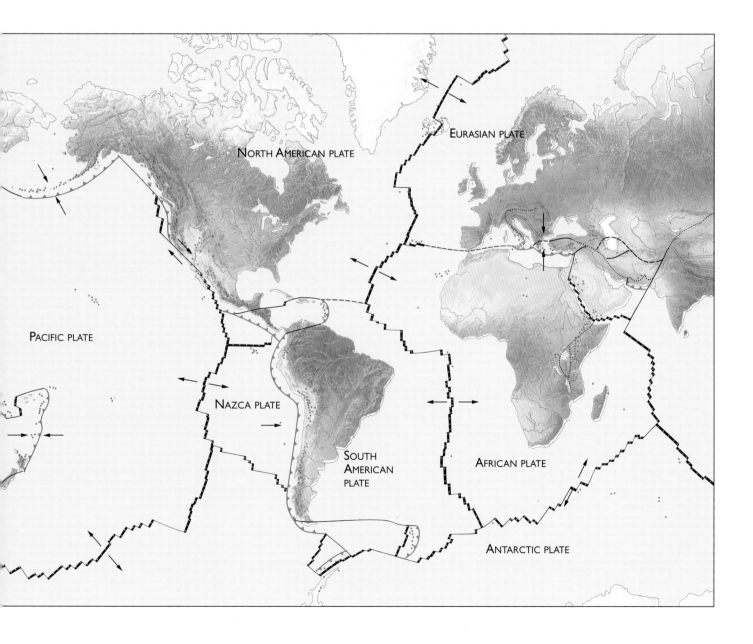

## PRINCIPLES OF PLATE MOTION

Rigid tectonic plates slide over the surface of the Earth, rotating about an axis called the Euler pole. The direction of relative motion between two plates (plates A and B) will always follow small circles around this axis. Transform faults, where the two plates are sliding past each other, will also follow these small circles, while mid-ocean ridges, where the plates are moving away from each other, usually follow great circles which pass through the Euler pole. The orientation of subduction zones, where the plates are converging, does not follow any general rule.

on a sphere in the 1700s. If two tectonic plates are truly rigid, then there will be a single Euler pole which defines the motion of any part of one plate relative to any part of the other plate.

Geologists have tried to see whether this is really the case. To illustrate how this is done, we can imagine that we have found an axis of rotation (Euler pole) which seems to work for the motion along some segments of the boundary between two plates. We then construct a map of the Earth, but with a difference. We make the top of our map the position of our estimated Euler pole for the two plates, instead of the true north pole that we would normally use. We draw on this map lines of longitude and latitude relative to our estimated Euler pole. If the tectonic plates are perfectly rigid, we will make an extraordinary discovery – the relative motion between the two plates will always be parallel to the latitude lines on our map. For example, transform

faults, where the two plates slide past each other, will plot as lines of latitude. But there is more. As the two plates move relative to each other, the crust breaks repeatedly in earthquakes. The horizontal slip during these earthquakes will also be parallel to our latitude lines. In other words, if the plates are rigid, the Euler pole makes sense of earthquakes. However, subduction zones, where the plates are converging, will not necessarily lie on lines of longitude or latitude on our map. Not all the Earth's surface is as rigid as plate tectonics demands – parts of the continents are not rigid at all (see Chapter 5). But leaving aside these relatively small regions, plate tectonics has proved to be a unifying concept which works extremely well for almost all of the ocean floor and much of the continents.

The speed at which the plates move relative to each other can now be measured directly by using an extremely precise surveying method called the global positioning system (GPS). This involves a network of satellites which are constantly orbiting the Earth, precisely tracked by radar stations. The distance between a ground station and the satellites is determined by the time taken for a high frequency radio pulse to travel between satellite and a receiver – sailors now routinely use a slightly less accurate version of this technique for navigation. By repeating these measurements over a period of years, it is possible to calculate the speeds of the tectonic plates. This work shows that plates are moving relative to each other at speeds up to 20 centimetres a year, and on average less than 10 centimetres a year. These may seem to be incredibly slow, but extrapolated over geological time the plates will travel substantial distances. The floor of the eastern Pacific Ocean is moving towards South America at about 9 centimetres per year. In the process, during the last 10 million years, the Pacific Ocean crust has slid under the western margin of South America and sunk nearly 1000 kilometres into the Earth's interior.

The moving plates carry the continents with them. If the history of the Earth over hundreds of millions of years is speeded up to take place over a few minutes, the continents appear to perform a strange dance as they drift across the surface of the Earth. In the past,

## CONTINENTS THROUGH TIME

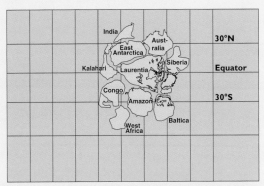

**(a)** Supercontinent of Rodinia 700 million years ago

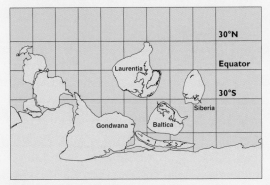

**(b)** Continents 500 million years ago, showing Gondwanaland near the south pole

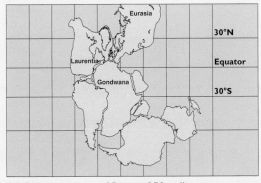

**(c)** Supercontinent of Pangea 250 million years ago

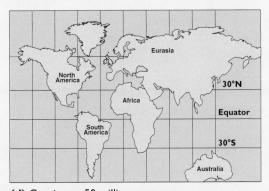

**(d)** Continents 50 million years ago

The continents are part of the great lithospheric plates which cover the surface of the Earth. As the plates move relative to each other, in the process of plate tectonics, the relative positions of the continents change. In the past, continents have split apart and collided, sometimes forming vast supercontinents.

supercontinents have split up, and smaller continents have come together to form new supercontinents. And as the continents have split, regrouped and drifted, they have had a major impact on both the global climate and life, constantly changing the face of the Earth (see Chapter 7).

The power of a theory is revealed when it can account for phenomena which were not part of the theory when it was first conceived. Like all good theories, plate tectonics is showing that power. For example, there is a simple explanation for the fact that the trenches and associated volcanic chains form such striking arcs. Try pushing in part of the surface of a ping-pong ball – the resultant depression is circular. Likewise, because the Earth is essentially spherical, wherever the surface bends down, as it does in subduction zones, a depression is created which tends to follow the arc of a circle. But the great chains of volcanoes which circle the Pacific are also arranged in a series of arcs, following precisely the arcs of trenches. This suggests that there must be a connection between the volcanoes and trenches. The volcanoes stand in line, like sentinels, above the dying oceanic plate, erupting a volcanic rock known by geologists as andesite after the volcanoes of the high Andes in South America.

## THE VOLCANOES OF THE HIGH ANDES

In Bolivia and northern Chile, a chain of volcanoes follows the western margin of the high Andes, rising from a base level of about 4000 metres to over 6000 metres. Many of these volcanoes are no longer active, but are the blasted and eroded stumps of volcanoes which erupted in the last 25 million years. But roughly every 50 kilometres along the volcanic chain, a conical volcanic mountain shows signs of life where a trail of white steam emerges near the summit. At the foot of these volcanoes hot water gushes out of cracks, leaving a white and yellow stain where minerals, leached from the volcanic rock, have precipitated in the dry and barren landscape.

It is usually possible to climb a volcano in a day if you are fit, scrambling up the steep scree slopes of volcanic ash. These lie between rocky ridges of lava, which has a distinctive composition of feldspar and other minerals. The lava is usually so sticky on eruption that it only flows a small way down the flanks of the volcano before solidifying. In this way, the steep slopes of the volcano have been built up, sometimes forming near-vertical cliffs.

As one climbs higher, a spectacular view of the distinctive volcanic landscape opens up. The volcanoes form conical mounds sticking out of a generally flat landscape mantled with volcanic ash. Much of the volcanic ash erupted as superheated dust clouds, at temperatures over 700°C. These ash clouds hug the ground, reaching speeds of 100 kilometres per hour and travelling for tens of kilometres over the surrounding landscape. As they travel, ash settles out and cools to become a hard compacted rock with a distinctive pale orange colour, known as ignimbrite. The ignimbrites have often overtopped low volcanic hills or climbed up the sides of the high volcanoes as they spread sideways from the site of the volcanic eruption. Where a volcano has blown out sideways in a particularly violent eruption, only the stump of the original conical volcano remains. Massive lumps of lava the size of houses lie scattered around. It was not until Mount St Helens erupted in May 1980, in what turned out to be the largest volcanic eruption in North

*Above left: A fumerole steams near the summit of an active volcano in the high Andes of Bolivia (Volcan Ollague). The volcano has steep sides because the lavas are too sticky to flow far. A black cinder cone, created during a volcanic eruption, is in the foreground.*

*Left: The high Andes in Bolivia are dotted with volcanoes, building up the land surface with volcanic rock. When viewed from space, cones of radially spreading lava are clearly visible – the summits of the volcanoes appear white if they are capped with snow. Volcanic ash mantles the ground between volcanoes.*

*Right: The summit craters of the active volcanoes in the high Andes of Bolivia are a steaming hell of hot gases such as water vapour, carbon dioxide and sulphur dioxide. The walls of the crater of Volcan Ollague are coated with yellow sulphur.*

## THE ROLE OF WATER AT A SUBDUCTION ZONE

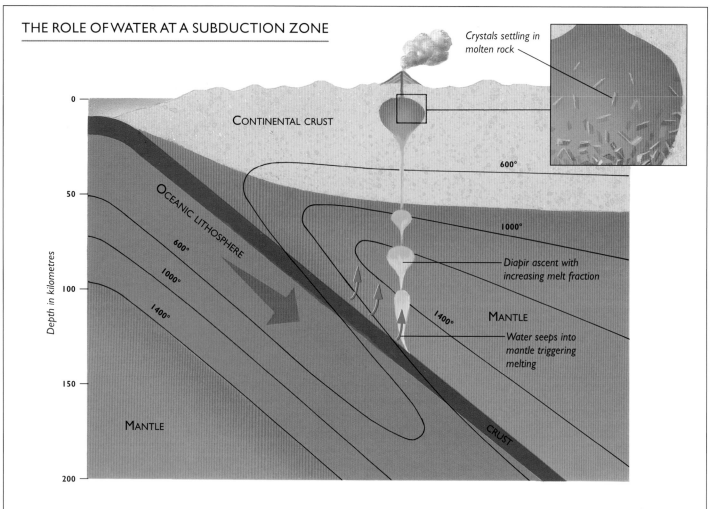

*Crystals settling in molten rock*

CONTINENTAL CRUST

600°

1000°

*Diapir ascent with increasing melt fraction*

OCEANIC LITHOSPHERE

600°

1000°

1400°

MANTLE

1400°

*Water seeps into mantle triggering melting*

MANTLE

CRUST

*Depth in kilometres*

0

50

100

150

200

When the oceanic lithosphere is subducted into the Earth's mantle, it carries down with it water which is locked up in the ocean crust. The water is eventually driven off by the high temperatures and pressure at depth, and the crust melts locally. The water

seeps into the overlying mantle, causing it to melt also. The molten rock collects together and, being less dense, rises up into the overlying cooler continental crust. Here it begins to crystallize. The early formed crystals are dense and settle to the bottom.

Some of the molten rock solidifies at depth to form large granite bodies. However, water-rich molten rock also reaches the surface, erupting explosively in the great chains of volcanoes which lie above the sinking ocean crust.

America during historic times, that anybody had ever recorded such a devastating sideways explosion.

As you approach the summit, the roar of the summit fumeroles where sulphur-rich steam gushes out at high pressure becomes ever more dominant. Here, the fumerole gas clouds have coated the volcanic rock with yellow sulphur and the air has a distinctive smell. A change in the wind direction can drown you in choking gas. Over the last few years, Leonore Hoke, an Austrian geologist who began her

research in Austria's eastern Alps, has sampled almost every volcanic fumerole in Bolivia, collecting the gas in specially designed copper tubes. She is interested in where the molten rock which feeds the volcanoes is coming from. Clues to this are found in the gases which are spewed out into the atmosphere. These gases include large quantities of water vapour, sulphur dioxide and carbon dioxide. In lesser quantities, but from Leonore Hoke's point of view equally as important, are nitrogen and helium.

## THE GAS FINGERPRINT

Helium is a peculiar gas. It is very light and also extremely unreactive. For these reasons helium is also very mobile. Anybody who has ever played with a helium-filled rubber balloon will know that eventually the balloon deflates and falls back to the ground. This is because the helium can escape through the rubber. Over a period of years, helium will even escape from a glass jar just by diffusing through the glass. As explained in Chapter 1, there are two forms of helium: a light form which has an atomic weight of three, i.e. three times as heavy as hydrogen, and a heavier form with an atomic weight of four. The latter is created when radioactive elements such as uranium and thorium decay, and is also called an alpha particle. Because the Earth's crust is enriched in uranium and thorium, it is constantly producing heavy helium. In fact, in average continental crust there is about 100 million times more heavy helium than light. At deeper levels in the Earth, below the crust in the mantle, the concentrations of uranium and thorium are much lower. Here, there is about 100,000 times more heavy helium than light. Put another way, the Earth's mantle is enriched a thousandfold, compared to the crust, in light helium. This makes it possible to trace the origin of any helium gas which is escaping from the Earth. If the ratio of helium-4 to helium-3 in the gas is much less than 100 million, say only one million, then you can be sure that at least some of the helium is coming from the mantle at depths of tens to hundreds of kilometres in the Earth.

Leonore Hoke has found that the helium in the gas escaping from the fumeroles on the high Andean volcanoes is extremely enriched in helium-3 – it contains up to 500 times as much as that in helium gas produced in the crust. This shows that some of the helium here originally came from the mantle, far below the crust. This tells us that at least some of the lava erupted in the volcano also originally came from the Earth's mantle. But the large quantities of water vapour in the lavas and volcanic gases suggest that this source region for the lava is also very wet. Other gases emitted by the volcanoes provide a clue to the source of this water. In particular, the gases are very rich in nitrogen, but contain only tiny amounts of argon and oxygen. This demonstrates that these gases are not mixed with air, because air is relatively rich in oxygen and argon. So if the nitrogen does not come from the air, where does it come from?

A high level of nitrogen is a common feature of gases emitted from volcanoes, not just in the Andes but throughout the great volcanic chains which encircle the Pacific, in New Zealand, Japan and Alaska. There is one abundant source of nitrogen not far from the volcanic arcs – the sediments at the bottom of the ocean. These sediments are full of the remains of dead organisms which once lived in the sea. Living organisms are rich in nitrogen – indeed we fertilize fields with nitrogen for this very reason to promote plant growth. The sea floor rocks are also saturated in water. These observations point to a surprising conclusion. Some of the gases coming out of the volcanic arcs come from the mantle, but some are coming from the sea floor. Armed with this observation, we are now ready to discuss a mechanism for the creation of the continental crust.

## SWANSONG OF AN OCEAN

If we recall our picture of the fate of the oceanic crust around the Pacific margin, we can imagine it bending down in the deep oceanic trench and diving down into the Earth's mantle, falling a few centimetres each year and triggering the swarm of earthquakes which make up the Benioff zones. The great chains of volcanoes virtually always lie between 100 and 200 kilometres vertically above the sinking plate. The proximity of the sinking plate to the volcanoes suggests that there must be some connection between them. Putting all the evidence together, geologists have constructed the following scenario for the origin of the volcanic arcs.

The ocean floor that returns to the Earth's mantle at a subduction zone is subtly different from the ocean floor that is created at the mid-ocean ridges. The deep sea black smokers and hydrothermal vents are evidence of the pervasive passage of hot water through the ocean crust. This water reacts strongly with the rock, changing its composition by adding salts and water and scavenging elements such as chrome and manganese. Eventually, as the ocean floor moves away from the mid-ocean ridge, a layer of deep ocean sediment is deposited on top, forming a new geological layer.

In a subduction zone, this altered ocean floor, often with part of its covering layer of sediment, is subjected to the high temperatures and intense pressures of the Earth's interior. For instance, at a depth of 50 kilometres, it will be at a temperature of hundreds of degrees centigrade, and squeezed under a massive pressure 15,000 times that on the Earth's surface. In response, changes occur in the rocks. Water and carbon dioxide, trapped in the rocks, are driven off. New minerals, stable at the higher temperature and pressure conditions, start to form. The intense shearing caused by the sliding action of the sinking slab may further raise its temperature. Eventually, parts of the ocean crust may become hot enough to melt. At depths in excess of 100 kilometres, minerals collapse under the immense pressures and more water is released from the sinking plate. The water, together with any molten oceanic crust, being lighter and more mobile than the surrounding rock, seeps out and rises into the overlying mantle rocks, which are at temperatures greater than 1000°C.

The large volumes of water squeezed out from the sinking oceanic crust have a curious effect on the surrounding hot mantle rocks. Laboratory experiments show that this effect is a bit like that of the antifreeze one puts in the car radiator, or the salt that is sprinkled on the roads in the winter. Both salt and antifreeze lower the freezing point of water. Put another way, if antifreeze or salt is mixed up with ice crystals at a temperature of a few degrees below 0°C, the ice crystals begin to melt. This is also exactly what happens when water penetrates the hot mantle rocks. The mantle is already close to melting, but the

addition of water is sufficient to make a small fraction (about 10 per cent) actually melt. Water vapour, carbon dioxide and other gases such as helium also separate out into the molten rock. This liquid, together with any molten oceanic crust, begins to pool and, being less dense than the surrounding rock, rises up vertically into the overlying crust to build the volcanoes, locally melting the crust on its way up (see p. 84).

## A GEOLOGICAL REFINERY

The study of subduction zones is helping to solve another puzzle about the Earth: why the Earth has two distinct types of crust – oceanic and continental crust. As we saw in Chapter 2, the oceanic crust has a very uniform thickness of about 7 kilometres and is mainly composed of the volcanic rock basalt. The continental crust has a different and much more variable composition and can be anything up to 80 kilometres thick. The existence of both oceanic and continental crust appears to be a unique feature of the Earth – the other terrestrial planets do not have such marked contrasts in the nature of their surface crust (see Chapter 8).

The continental crust has a composition similar to that of lavas erupted in volcanic arcs, and its origin is inextricably tied up with these volcanoes. But the oceans themselves play a crucial role. To understand why, we need to investigate in more detail the upward journey of the molten rock which feeds the volcanoes as it rises from its source in the mantle. It has a similar composition to the ocean crust – this always seems to be the composition of molten rock generated in the mantle – but it will be rich in water (about 3 per cent by weight). But both volcanoes at the surface and the continental crust have a different composition, being roughly three parts basalt and one part granite. The problem is to understand how the one part granite is produced.

Another way of expressing this is in terms of the content of silica (silicon dioxide), which is an

*Granite is a common rock type in the continents. When viewed under the microscope, granite appears as a mosaic of different crystals such as quartz and feldspar (white to blue) and mica (brown, yellow, pink and green). The field of view is about one centimetre across.*

important component of all rocks. Basalt is about 50 per cent by weight silica, but the continental crust contains much more, being on average 60 per cent silica (granite is about 75 per cent silica). Thus the creation of continental crust involves some sort of refining process which increases the concentration of silica. This must go on as the molten rock rises up to the surface through the crust of the overlying plate and cools. Geologists now have a good understanding of what happens, based on many experiments on molten rock.

Deep in the crust, the molten rock solidifies slowly as the minerals, which make up the rock, crystallize. But the full range of minerals does not crystallize all at once. Instead, they crystallize in a particular order. The early-formed crystals are those which have the

highest melting points – these are magnesium-rich minerals such as olivine and pyroxene. Being dense, they tend to sink to the bottom of the body of molten rock, leaving behind a liquid which is relatively poor in magnesium but, importantly, enriched in silica (see p. 84). As this goes on, the silica content of the remaining liquid progressively increases, approaching the composition of the continental crust. Water plays an important part in this, because the water promotes the crystallization of minerals typically found in the continental crust. It also keeps the rock molten at lower temperatures than would otherwise be the case. Thus, the refining process can go on for longer, further enriching the remaining molten rock in silica. Eventually, the main components of the continental crust – quartz, mica and feldspar – crystallize.

The molten rock reaches the surface as a sticky lava called andesite, which has a composition similar to that of the continental crust. Not all lavas erupting from the volcanoes have this composition. Sometimes the molten rock, with its original basaltic composition, makes it all the way to the surface, building black cinder cones. Sometimes the refining process, during rock crystallization, results in a liquid which is unusually enriched in silica and more sticky than normal – the stickiness is largely determined by the silica concentration. This may cool at depth as vast granite bodies, or it may ooze out at the surface like toothpaste, forming broad domes of lava. Sometimes there is so much water vapour trapped in the lava that it explodes, sweeping out sideways from the volcano or creating vast ash clouds above the volcano which eventually cover the landscape with superheated ash.

## BUILDING THE CRUST

In the volcanic arcs, we are witnessing a process of differentiation in which the crust separates out from the mantle. Geologists reckon that the molten rock, eventually added to the crust, travels on average at about one metre each year, taking a few hundred thousand years to reach the surface from its source region in the mantle. This is a one-way road, because the molten rock is less dense than the solid mantle and remains on top, essentially floating. Slowly, over geological time, it builds up the continental crust. There is considerable debate among geologists about the exact rate at which this has happened in the past. There may have been periods in the Earth's history when continents were growing more rapidly. Some geologists argue that at least half of today's continental crust already existed about 2.5 billion years ago. However, estimates of the volume of volcanic rock in the Bolivian Andes suggest that, over the last 50 million years, the crust beneath the volcanic arc has become one and a half times thicker because of the accumulation of molten rock. The

conclusion is clear: continents are being created today, possibly at a rate of about one cubic kilometre every year.

Sinking oceanic plates do not build the continents solely by generating molten rock. At the deep sea trenches, where the ocean floor starts to bend down into the Earth's interior, the continents are also growing. As the ocean floor slides underneath the edge of the adjacent crust, slivers of ocean floor rocks are scraped off. These slivers stack up on one side of the trench (see pp. 76–7). Occasionally, a volcanic island arc is dragged into a trench. This too is plastered to the trench wall, adding to the flotsam and jetsam of old oceans accumulating here, along with sediment eroded from the adjacent landmass. Large tracts of the continents, bordering the Pacific Ocean, have been built up in this way. The subduction zone is behaving rather like the skimmer box or filter in a swimming pool. The tectonic plate slides sideways towards the trench, just like the surface of the swimming pool as it is sucked towards the skimmer box. And like the leaves floating in the pool, caught by the skimmer box, lighter or high standing parts of the oceanic plate are trapped at the trench, scraped up and amalgamated to form the roots of new continents.

Sometimes, a rather curious process occurs along the volcanic arc. The crust splits open to form a rift in the Earth's surface (see pp. 76–7). If this continues, a narrow ocean basin is created, complete with its own mid-ocean ridge. The traces of these back-arc basins, as they are called, can be seen around much of the western Pacific Ocean, extending from the North Island of New Zealand northwards just to the west of the Kermadec, Tongan and Fijian islands. Intricate swirling patterns of island arcs and back-arc basins create a tangle of small ocean basins between the Solomon Islands and New Guinea, and among the spice islands of the East Indies. The Sea of Japan, which separates Japan from the mainland of Asia, is another. Geologists have now dated the ocean floor in the back-arc basins, using the characteristic pattern of magnetic anomalies, or the age of fossil planktonic creatures preserved on the sea floor. These show that most of the back-arc basins have formed in the last few tens of millions of years and are young features of the Earth's surface. They seem to form because the place where the ocean floor dives back into the Earth's interior, at the ocean trench, does not stay fixed in space, but also migrates across the surface of the Earth. If the trench moves away from the overriding plate, a new ocean basin opens up behind it.

To geologists, subduction is a wondrous process, revealing a remarkable economy of Nature. When oceans die, new continental crust is created. Even as geologists investigate the giant volcanoes in the world's volcanic arcs, this is going on beneath their feet. The new crust, because it is less dense than oceanic crust, sits high on the Earth's surface, forming the dry land. We live on this dry land. We benefit in other ways as well.

## A RICH DIVIDEND

As a coda to the story of subduction, we return to the subject of that mineral wealth – copper, tin, lead, silver and gold – which attracted many of the early explorers to the Ring of Fire in the first place. It is now clear that the storehouse of riches is in both the Earth's mantle and the sinking ocean crust. Indeed, rich mineral deposits lie on the ocean floor today at the foot of the black smokers (see Chapter 2). The metals are carried up with the molten rock that feeds the volcanoes in the volcanic arcs. But they need to be concentrated before they will form a rich vein worth mining. How this happens was graphically demonstrated when a geothermal power station in New Zealand began to clog up.

In the 1970s, when oil prices were high, the New Zealand government was investing heavily in geothermal power. They had built a power station at Wairakei, right among the active volcanoes in the North Island. These volcanoes, like the volcanic chains in the rest of the Pacific margin, lie about 100 kilometres vertically above the sinking Pacific Ocean crust. The power station was exploiting the abundant hot water which emerges in geysers or bubbling mud

pools in the area. The engineers drilled deep holes into the hot ground and pumped the hot water into heat exchangers which drive large steam turbines. The flow in each pipe was controlled by a heavy duty valve. The power station operators found that over time it became more and more difficult to control the flow through these valves. Eventually they dismantled the valves and to their surprise found that the inside of each valve was encrusted with an alloy of gold and silver called electrum.

It became clear that the superheated water was boiling in the valves, releasing hydrogen sulphide gas with its characteristic rotten egg smell. The precious metals were being carried together with sulphur in the water as a complex molecule. The release of hydrogen sulphide broke up this molecule and gold and silver precipitated in the valve. Eventually, as large quantities of water passed through the valve, enough material precipitated to clog up the power station. When this discovery was eventually published, there was enormous interest from mining companies. The research suggests that wherever there are boiling hot springs and geysers which smell of hydrogen sulphide, precious minerals are likely to be accumulating not far away.

A more significant discovery has been that the relatively small volumes of water handled by a geothermal power station in a volcanic arc, can carry enough metal to create a mineral deposit. Therefore, over geological time, fabulously rich deposits can form at shallow levels in the volcanic chains. Here is the clue to the origin of the Cerro Rico in Bolivia with its rich silver veins. It was a volcano, active some 16 million years ago, through which large volumes of boiling water gushed, driven by the heat from molten rock. At that time, it stood above a subduction zone where a much older piece of the Pacific oceanic crust was sinking back into the Earth's interior. Its mineral wealth is part and parcel of the process of building up the continental crust. However, its extreme richness must still remain a quirk of nature. Perhaps the native Bolivians, who still work in the silver mines of the Cerro Rico, are right after all when they see this as a gift from Pachamama, the Indian god of the Earth's interior.

# CHAPTER 4

# A VOLCANIC IMAGINATION

· · · · · · · · · · · · · · · · · · · · · · · · · · · · ·

What drives the tectonic plates as they glide
over the Earth's surface? Searching for an answer,
scientists have probed our planet to its core. In this
realm of unimaginably high temperatures and
pressures, matter takes on new forms, and
solid rock can behave like a fluid. As vast masses
of rock flow slowly within the Earth, so the surface
moves and changes. Gigantic plumes of hot material
can well up from the depths, triggering huge volcanic
eruptions and causing the crust to bulge and break.
The result may be the splitting of a continent
and the creation of a new ocean basin.

*A splash of lava throws up droplets of liquid rock into the air as a river of molten
basalt flows rapidly to the sea during a volcanic eruption in Hawaii.*

## A SPINNING DROPLET

The interior of the Earth has always been fair game for speculation. It is way beyond our reach, and yet – as we shall discover later in this chapter – virtually every aspect of the planet's surface is ultimately connected with what happens here. Scientists have persuaded governments to spend tens or even hundreds of millions of pounds drilling superdeep holes into the Earth's crust, pushing this technology to its limits. But these holes have no more than scratched the surface, penetrating only a paltry 12 kilometres or so.

In 1864, Jules Verne wrote his classic book *Journey to the Centre of the Earth*. He indulged the fantasy of many an armchair geologist by allowing his heroes to actually make a journey deep inside the Earth. Near the beginning of the book, the two main characters – Professor Lidenbrock and his nephew Axel – discuss the feasibility of such a journey. They argue whether the centre is liquid, solid or gas (our italics):

*Axel*: 'It is generally recognized that the temperature rises about one degree [*Fahrenheit*] for every seventy feet below the surface [*roughly 25°C per kilometre*]; so that if you admit that ratio to be constant, the radius of the Earth being over four thousand miles, the temperature at the centre must be over two million degrees [*unfortunately Axel has made a small mathematical slip in the heat of the moment – he meant to say 300,000 degrees, though this is still a very high temperature*]. Consequently all the substances inside the earth must be in a state of incandescent gas…'

*Prof. Lidenbrock*: 'Neither you nor anybody else knows for certain what is going on inside the Earth, seeing that we have penetrated only about one twelve-thousandth part of its radius….let me tell you that some real scientists, including Poisson, have proved that if a temperature of two million degrees existed inside the globe, the fiery gases given off by the melted matter would acquire such elasticity that the Earth's crust would be unable to resist it, and that it would explode like the plates of a bursting boiler.'

*Axel*: 'That is Poisson's opinion, Uncle, nothing more.'

*Prof. Lidenbrock*: '[*Humphry Davy and I*] spent a long time discussing the hypothesis of the liquid nature of the terrestrial nucleus. We were agreed that this liquidity could not exist…Because this liquid mass would be subject, like the sea, to the attraction of the Moon, and consequently, twice a day, there would be internal tides which, pushing up the Earth's crust, would cause periodical earthquakes.'

Though this dialogue is taken from an adventure story, it illustrates beautifully how scientists have had to go about exploring the centre of the Earth: by inference and enormous extrapolation from what they can see at the surface. It is rather like trying to work out the contents of a mysterious parcel just by examining the outside, perhaps feeling its weight, shaking it and scrutinizing in minute detail the wrapping paper. Proceeding in this spirit, the chapter is a sort of quest, following up the clues which help us work out what is inside the Earth. The holy grail of this quest is nothing less than an explanation for the geological activity we see on the surface, which manifests itself as the motions of the tectonic plates and volcanoes. We can begin with a couple of simple observations.

*First observation*: On a ship in the middle of the ocean, you will see nothing in all directions except a watery horizon. Sailors have long known that even from the top of a high mast, this horizon is only about 25 kilometres away. Beyond that you cannot see because of the curvature of the Earth. In about 200 BC, the ancient Greeks used this curvature to calculate the basic dimensions of our planet. The spinning ball we live on is about 12,700 kilometres in diameter and about 40,000 kilometres in circumference.

*The setting Sun, as it sinks below the horizon at sea, is clear evidence that the Earth is not flat, but a sphere. Because of the curvature of the Earth's surface, one cannot see beyond the horizon.*

*Second observation:* An object falls down if left unsupported. Isaac Newton showed that this was because of the force of gravity between the mass of the object and the mass of the Earth. He also discovered that we can measure the mass of the Earth by observing the force of gravity at the Earth's surface. The mass of the Earth turns out to be roughly 6000 million million million tonnes.

To begin with, we can deduce the density of the Earth – this is the mass divided by the volume and comes to five and a half times the density of water. We can immediately say that the interior cannot be the same stuff as the surface, because the density of almost all surface rocks is less than three times that of water.

A closer scrutiny of the shape of the Earth is also revealing. The Earth is not a perfect sphere, but bulges at the equator to form a slightly flattened ellipsoid (see p.102). By the mid 1700s, physicists had already shown that the shape of the Earth is almost precisely what one would expect for a massive spinning fluid droplet. The force of gravity draws the fluid mass of the planet together. The fluid is spread out slightly at the equator because of the rotation of the Earth, and is slightly flattened at the poles. But this fluid must be a very stiff fluid because, as Professor Lidenbrock in Jules Verne's book rightly points out, if it was as runny as molten rock, there would be a large tidal effect. In fact, there would be a very different sea tide, because the sea and Earth would be pulled more equally by the Moon.

## LOOKING INSIDE

There comes a point in the game of working out the contents of a mysterious parcel when one starts tapping it in the hope of using one's sense of touch or hearing to reveal something. Geologists have employed virtually the same technique for probing the Earth. But in comparison to our parcel, the Earth is huge. The sort of tapping which has any hope of penetrating the vast bulk of the Earth is generated when a portion suddenly fractures, triggering an earthquake. A large earthquake releases in the form of seismic vibrations more energy than a 100-megaton bomb (*seism* means shock). These vibrations form waves which spread out from the source of the earthquake, rather like the ripples in a pond created when a stone is dropped in.

Here, we need a brief digression on waves or vibrations in general. Any vibration has a frequency. This is merely the number of times a second that the vibration repeats itself, like the swings of a clock pendulum. For example, deep traffic rumble is barely audible but can be felt, vibrating at about ten cycles per second. Another way of describing this is in terms of the time taken for a single vibration or complete swing of the pendulum – this is the period of the vibration. Traffic rumble has a period of about a tenth of a second. Earthquake vibrations generally have much longer periods, in the range of one second to several hours, and are at too low a frequency to be heard with the human ear.

An earthquake sets the Earth vibrating in a number of different ways. The whole Earth starts 'ringing', vibrating with a period of about an hour. Waves with periods between ten and several hundred seconds travel along the surface of the Earth and cause the characteristic ground shaking (surface waves). Shorter period vibrations (about one second) travel through the interior of the Earth, forming 'body' waves. It is the latter that tell us most about the internal structure of the Earth. These can be detected with a seismometer, which is not much more than a carefully

balanced weight, suspended by a spring and mounted on solid rock (see opposite). The spring starts to vibrate as it is shaken by the earthquake. With old-fashioned instruments, a pen mounted on the end of the spring traces out a wave-like pattern on a rotating drum of paper, generating a seismogram. In modern instruments, the vibrations are electronically recorded and stored digitally in a computer.

A typical seismogram for an earthquake consists of a series of squiggles of varying amplitude. The body waves arrive first. The initial train of vibrations felt by the instrument are called, unsurprisingly, primary waves or P waves for short. These waves, in fact, are very similar to sound waves; the ground oscillates back and forth, like a vibrating spring, in the direction the wave is travelling. P waves arrive first because they travel the fastest of all the vibrations triggered by the earthquake. Next to arrive at the seismometer are the secondary waves (S waves). These travel slightly slower than the P waves, and vibrate at right angles to their direction of motion, more like the motion of a snake as it winds its way over the ground. After the short period P and S body waves, the slower and longer period surface waves are eventually picked up by the seismometer. These cause a much more violent shaking of the instrument than the body waves, and can easily push the pen recorder off the scale.

If we recall our analogy of ripples in a pond, triggered as a stone hits the water, we can imagine the ripples moving outwards in concentric circles away from the source of the disturbance. But rather than thinking about circles of ripples, we can think instead of rays of energy spreading out radially in all directions from a common centre. Thus, during an earthquake, there will be rays of P waves and S waves moving outwards away from the source of the earthquake and penetrating into the Earth. The theory of this ray propagation has been well worked out by seismologists. The precise speed which these rays will travel depends on both the strength and density of the rock through which they pass. The rays may be bent or deflected if the strength and density change. For example, a ray which leaves the earthquake source, pointing down, may curve upwards to reach the surface at a point further round the Earth from the

# SEISMIC WAVES

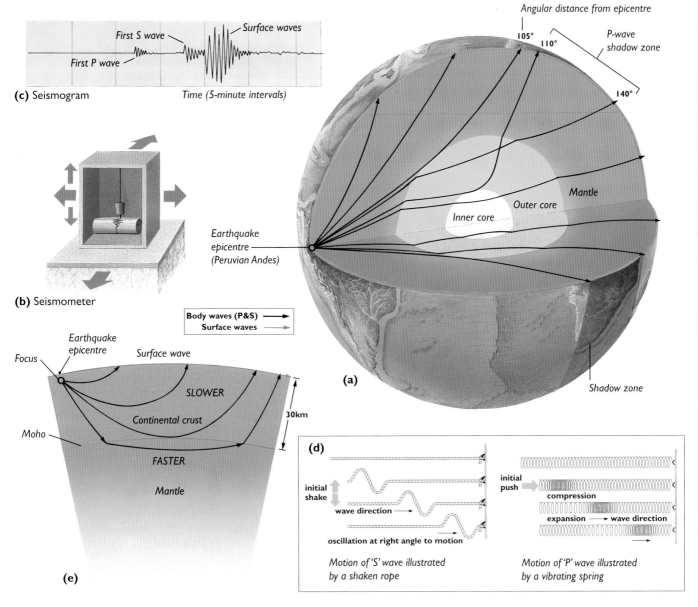

**(c) Seismogram**

First S wave
First P wave
Surface waves
Time (5-minute intervals)

**(b) Seismometer**

Angular distance from epicentre
P-wave shadow zone
105°
110°
140°
Mantle
Outer core
Inner core
Earthquake epicentre (Peruvian Andes)
Shadow zone
**(a)**

Body waves (P&S)
Surface waves

Focus
Earthquake epicentre
Surface wave
SLOWER
Continental crust
30km
Moho
FASTER
Mantle
**(e)**

**(d)**
initial shake
wave direction
oscillation at right angle to motion
*Motion of 'S' wave illustrated by a shaken rope*

initial push
compression
expansion → wave direction
*Motion of 'P' wave illustrated by a vibrating spring*

*Sensitive seismometers, mounted on slabs of concrete at Eskdalemuir in Scotland, form part of a global network (Worldwide Standardized Seismography Network) set up in the early 1960s to detect earthquake vibrations.*

Earthquake vibrations can be used to probe the interior of the Earth. The vibrations radiate from the earthquake as rays which travel through the Earth's interior **(a)**, before reaching the surface where they are detected with a sensitive instrument called a seismometer **(b)**. The seismometer records the vibrations as a characteristic pattern of wiggles called a seismogram **(c)**. This shows that vibrations, called P waves, which have travelled through the Earth's interior, arrive first; P waves are followed by the slightly slower S waves. Subsequently, waves which have travelled over the surface of the Earth are picked up. P waves are vibrations rather like the oscillations of a spring, while S waves have a characteristic sideways motion **(d)**. The arrival times of earthquake waves show that the Earth's interior consists of an outermost crust **(e)**, overlying the mantle and core. The outer core is liquid – S waves cannot travel through this region and P waves are bent and slowed down, resulting in a marked shadow zone on the other side of the Earth to that of the earthquake, where P waves are poorly detected.

earthquake. Since the turn of the century, the arrival times of these rays at various points on the Earth's surface have been recorded for many earthquakes and have been published as 'Travel-Time' tables, thanks to the colossal efforts of three pioneering seismologists – Beno Gutenberg, Sir Harold Jeffreys and Keith Bullen.

The 'Travel-Time' tables are a sort of Rosetta Stone for seismologists, providing the key to looking inside the Earth. But to use them, seismologists have to work backwards, as it were, trying to deduce what the Earth is like from the time taken for the rays to pass through it. The analysis is usually done by assuming that the Earth consists of a number of spherical shells in which the waves travel at different speeds. It is rather like trying to work out what combination of bus, train and walking, each with its characteristic speed, somebody used to travel to work, knowing only the total time and distance of the journey.

The Earth viewed through the medium of earthquakes looks rather like an onion. We have encountered the outermost layer, called the crust, in previous chapters. Compared to the radius of the Earth (about 6400 kilometres), this layer is like the thin papery skin of the onion, being only about 7 kilometres thick beneath the oceans, and up to 80 kilometres thick in the continents. Both P and S waves travel relatively slowly in the crust; a P wave speed of about 6 kilometres per second is typical. The base of the crust is called the Mohorovičić discontinuity, or Moho for short, after the Croatian seismologist who first discovered it in 1909. Below the crust, the body waves travel progressively faster down to a depth of about 2900 kilometres. This high speed zone in the Earth is the mantle; P wave speeds increase from about 8 kilometres per second at the top of the mantle to nearly 14 kilometres per second at the base. But there is internal layering within the mantle. For example, a zone with slightly lower velocities is found beneath the oceans, starting at depths of a few tens of kilometres. This may mark the bottom of the lithosphere and top of the asthenosphere. Between depths of 400 and 700 kilometres, both P and S wave speeds increase rapidly. This change is often used to divide the mantle into an upper and lower part.

# The interior of the Earth

The Earth has an onion-like structure, varying in composition and material properties – this is revealed by the time taken for vibrations from earthquakes to travel through the Earth's interior. Seismic waves travel at progressively faster speeds in the solid mantle. Beneath the mantle is the core; we know the outer core is liquid because seismic waves either cannot pass through it, or travel more slowly there than in the mantle. Both the inner and outer parts of the core consist of iron and nickel – these have a higher density than the rocks in the mantle.

Complex flow in the liquid outer core generates the Earth's magnetic field. The overlying solid mantle is also in constant motion – the outermost part, together with the crust, forms rigid lithospheric plates which glide over the underlying weaker parts of the mantle.

The oceanic lithosphere is created along the axis of the mid-ocean ridge where the mantle rises close to the surface. At subduction zones, the plates sink down deep into the mantle. Hot plumes rise up from near the base of the mantle, triggering volcanic activity on the surface. The rising and sinking regions create a pattern of convection which carries heat from the Earth's interior to nearer the surface.

## Section through Earth

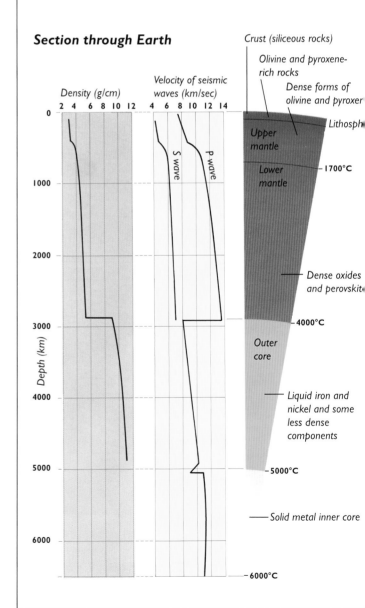

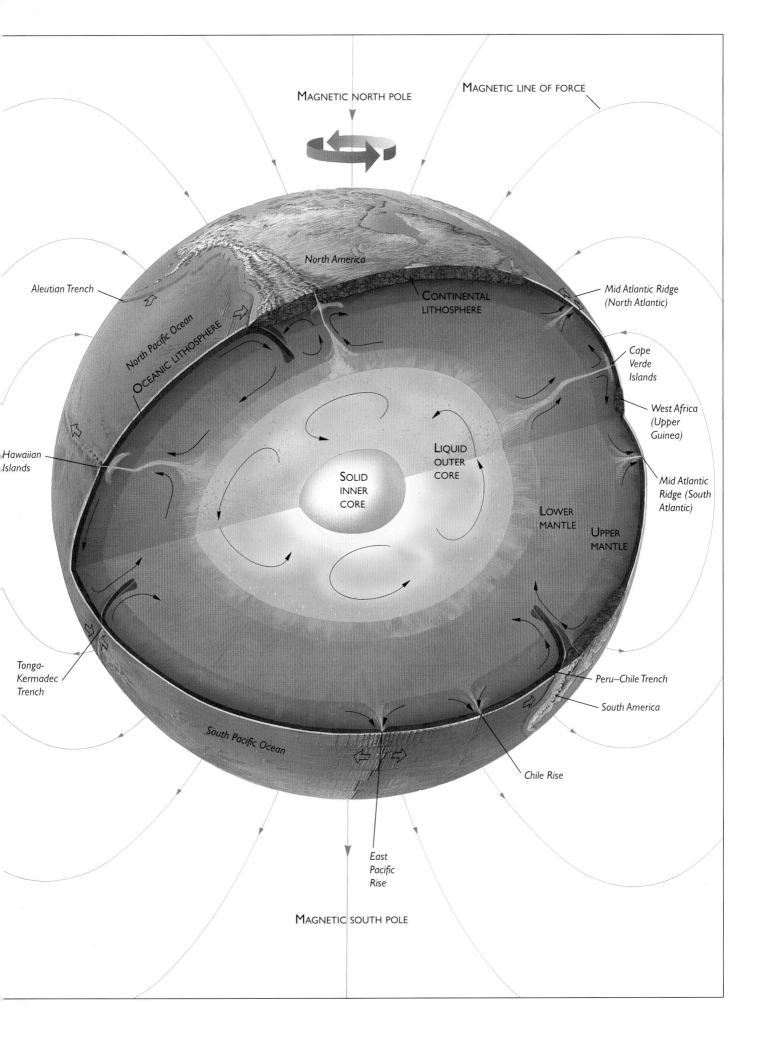

MAGNETIC NORTH POLE

MAGNETIC LINE OF FORCE

North America

CONTINENTAL
LITHOSPHERE

Aleutian Trench

Mid Atlantic Ridge
(North Atlantic)

North Pacific Ocean

OCEANIC LITHOSPHERE

Cape
Verde
Islands

West Africa
(Upper
Guinea)

LIQUID
OUTER
CORE

Hawaiian
Islands

SOLID
INNER
CORE

Mid Atlantic
Ridge (South
Atlantic)

LOWER
MANTLE

UPPER
MANTLE

Tonga-
Kermadec
Trench

Peru–Chile Trench

South America

South Pacific Ocean

Chile Rise

East
Pacific
Rise

MAGNETIC SOUTH POLE

Curiously enough, the centre of the Earth, forming that bit of the onion enclosed by the mantle layer, is the easiest to recognize with earthquakes and was first discovered in 1897. Imagine an earthquake as a sort of light bulb, positioned on one side of the planet, radiating out rays of seismic 'light' which travel through the Earth's interior. The vibrations felt over the Earth's surface suggest that something inside the Earth is behaving just like a magnifying glass, focusing the P wave rays into a seismic bright spot on the other side of the Earth, diametrically opposite to the earthquake. Around this bright spot, there is a dark ring where the expected P waves do not arrive. The object inside the Earth which is deflecting the earthquake energy is the core. It does this by slowing down the speed of the P waves.

So far we have discussed the internal structure of the Earth only in terms of the speed of the seismic body waves. But we can in fact make some very definite statements about the Earth's interior without worrying too much about the precise speeds. Firstly, S waves can only pass through solid material. So the mere fact that S waves – vibrating with a snake-like sideways motion – travel through the Earth's crust and mantle show that this portion of the Earth is solid, though there might be very small pockets of molten rock. In fact, the mantle must have the strength of steel to explain the speed at which the S waves travel. There seems to be a contradiction here: in contrast to the high strength suggested by the earthquakes, we have already described evidence at the beginning of this chapter that the mantle behaves like a fluid – we will return to this later. However, S waves do not travel through the core, and the P waves travel quite slowly here. This shows that at least part of the core is liquid. A more detailed analysis of the travel times of P waves indicate another layer within the core, at a depth of about 5000 kilometres, is in fact solid. So the core has an outer liquid portion and an inner solid part. Combining the information from P waves and S waves, seismologists have estimated some of the properties of the different layers; for example, the core is much denser than the rest of the Earth. These densities have turned out to be an important clue to working out what is actually inside the Earth.

## WHAT IS THE EARTH MADE OF?

This is an easy question to answer for the surface of the Earth, which forms the crust – we have discussed the composition of the oceanic and continental crust in Chapters 2 and 3. But what about below this, in the mantle?

Occasionally, volcanic eruptions tap deeper parts of the Earth. These eruptions have interested people other than geologists because they contain diamonds. The diamonds occur as rare plums in a pudding of rock, but they provide a clue to the depth in the Earth from which the eruptions come. This is because diamonds are a form of carbon which can only be made at very great pressures – the pressure inside the Earth increases with depth because of the weight of the overlying rock. Some of these diamond-bearing eruptions are estimated to have come from depths of 150 to 300 kilometres.

The rock pudding in which the diamonds sit is often made of the minerals olivine, garnet and pyroxene. Rocks of this composition, when they melt, produce a molten lava which geologists call basalt. Basalt is the material which is erupted on the sea floor at the mid-ocean ridges and also in many volcanoes on land. This suggests that the Earth's mantle, at depths down to several hundred kilometres, is mainly made of the minerals olivine, garnet and pyroxene (often forming a rock called peridotite), a composition very different to the crust.

No samples of the mantle deeper than several hundred kilometres have ever been obtained. So scientists enter a realm of extrapolation when talking about the composition of the Earth at these greater depths. One argument is that if you know the original bulk composition of the Earth when it formed as a

*Right: The upper part of the Earth's mantle is made of a rock called peridotite. When viewed under the microscope, individual crystals of olivine are visible, coloured yellow, pink, green and blue with criss-crossing fractures. The field of view is about 3 millimetres across.*

planet, then by looking at the composition of the layers nearer the surface, you can work out what is missing. The 'missing' material must be deeper inside the Earth. For a whole variety of reasons, which we discuss in Chapter 8, most geologists believe that fragments of rock which drift in space, occasionally falling to Earth in meteorite showers, are pieces of the material from which the Earth coalesced during the creation of the Solar System.

The commonest type of meteorite, called a chondrite (see p. 210), is predominantly composed of four elements – iron, oxygen, magnesium and silicon. Another common meteorite consists of an iron–nickel alloy. If the Earth has the same composition as a chondrite, but with most of the iron in the same metallic form as in the iron–nickel meteorites, then this would produce a planet with a similar density to the Earth's. But there is more. A chondritic meteorite minus its iron has a similar composition to the minerals olivine and pyroxene, which certainly exist in the top part of the mantle. The amount of iron left over, in metallic form, will neatly fit into the core. In addition, the density of the Earth's core is so high – about twelve times that of water – that it is hard to explain unless the core is mainly iron. Thus, surprising as it may seem, it is possible by combining both the structure of the Earth, deduced from earthquakes, with the known composition of meteorites to estimate fairly well the composition of the Earth's interior. And one of the principal conclusions is that the central part of the Earth, occupying a region nearly the size of Mars, is made mainly of iron and much of this molten.

There is a slight catch here. Knowing the composition of a material is not necessarily enough to tell you what it is. Under the intense temperatures and pressures of the planet's interior, minerals can be transformed into new and exotic forms. For example, coal is mainly carbon. But carbon, when subjected to conditions deep within the Earth, is transformed into diamond, a material with very different properties to coal. So to discover how the minerals that make up the mantle and core actually behave inside the Earth, scientists have had to simulate this region in the laboratory.

## AT THE TIP OF A DIAMOND

Andrew Jephcoat runs a laboratory at Oxford University which is dedicated to reproducing conditions deep within the Earth. To do this, he occasionally visits Amsterdam to buy gems. In fact, Jephcoat is after high quality diamonds. These diamonds are the key to his method of working. Diamonds are extremely strong and capable of withstanding enormous pressures without fracturing. They are also transparent. The method Jephcoat uses is in principle very simple – though in practice there are many technical problems which have to be overcome.

The diamonds are cut into cones and arranged so that two of them meet at the tips of the cones. They are then placed in a heavy steel cylinder. Plungers at either end of the cylinder push the diamonds together, and the whole apparatus is placed between the jaws of a hydraulic press. The experiment takes place between the tips of the diamond cones, and can be viewed with a microscope right down the transparent cone axis. During an experimental run, a piece of the Earth's mantle is placed between the tips of the diamond cones, and squeezed together with the hydraulic press. The diamond tips transmit the force applied by the press but, because this is concentrated into a microscopic area, the pressure is enormous and approaches that at hundreds or thousands of kilometres in the Earth. To obtain the possible temperatures at these depths, a laser beam is fed in through a hole to the diamond tip. The beam is highly focused so that temperatures of thousands of degrees can easily be reached. When the sample has reached the target conditions various physical properties are measured, such as the speed of seismic vibrations passing through the sample, as well as the electrical resistance and thermal conductivity.

The high pressures dictate the crystal structure of the materials inside the Earth. The experiments show that the minerals olivine and pyroxene, which both exist in the the upper part of the mantle, transform at greater depths into more compact and dense minerals. At depths greater than about 700 kilometres, in the lower mantle, the minerals change into a very tight-packed structure called perovskite – this is probably the most abundant mineral on Earth, but exists on the surface only in minuscule quantities between the tips of a diamond anvil.

Laboratory experiments have also revealed the temperature near the centre of the Earth in the core. As we have seen, seismic studies show that the outer core is liquid, almost certainly molten nickel and iron. The inner core is assumed to be solid nickel and iron. So the temperature at the interface between the liquid outer core and solid inner core is the melting point of iron and nickel. This melting point is not what would be measured at the Earth's surface, but is much higher because of the huge pressure in the core. So, simply by finding the temperature at which iron and nickel melt at this pressure, produced momentarily in the laboratory, it is possible to discover the temperature near the centre of the Earth. This turns out to be a colossal 5000°C – nearly the same as the surface temperature of the Sun.

A molten outer core, consisting mainly of iron and nickel, may act as the dynamo which generates the Earth's magnetic field (see Chapter 2). This dynamo cannot be a simple one because iron at core temperatures would not normally be expected to show any magnetic properties. However, iron is a good electrical conductor. Physicists have now worked out how a complicated pattern of flow in the core, creating electrical currents, could set up a magnetic field. This dynamo seems to be inherently unstable – not only does the magnetic field change on a yearly basis, but every million years or so it flips altogether, so that a compass needle would point to the south rather than the north pole (see Chapter 2).

## GOING WITH THE FLOW

Our investigation of the Earth's interior has given us an idea of the material with which it is constructed. If we want to make further progress in the quest we set ourselves at the beginning of this chapter, we now have to face up to a paradox about the strength of this material, which we have so far glossed over. Earthquake vibrations suggest that the mantle is as strong as steel. But the shape of the Earth suggests that it is a spinning fluid droplet, with virtually no long-term strength at all. If we look yet more closely at its shape, there is further evidence for its fluidity.

There are dents and bulges in the surface around Scandinavia and northern Canada, where it slightly deviates from the droplet shape. Here, around the margins of both the Baltic Sea and Hudson Bay, there are the remains of beaches, hundreds of metres above sea level. In the early eighteenth century, a Swedish scientist, Anders Celsius, proposed that the old stranded beaches traced out a shoreline when sea level was much higher than today. Celsius even went so far as to place a mark on a rock so that he could monitor the subsequent year by year 'drop' in sea level below the mark. It appeared to Celsius that sea level was falling by about 3 centimetres a year. But Celsius could not be sure that it was, in fact, the *sea* which was falling, rather than the *land* rising. The latter would indicate vertical movements in the Earth's interior.

Changes in sea level have a very practical significance: they cause problems for harbours, which require a certain depth of water for shipping. Almost all harbours around the world now routinely monitor sea level with a tide gauge. This is a float which sinks and rises with the tides. The vertical movement of the float is monitored by a chart recorder, so that over a period of months or years a wave-like pattern of the rise and fall of the tide is produced. If sea level is constant, then the mean tide level over a period of years will lie on a horizontal line in the record. If sea level is changing, then this line will slope. Tide gauges throughout the world generally show very little movement of sea level. But the tide gauges in Scandinavia and Hudson Bay show a marked change in sea level each year, similar to that found by Celsius. The only possible explanation is that global sea level is a more or less constant level, but the land in Scandinavia and the Hudson Bay region is rising.

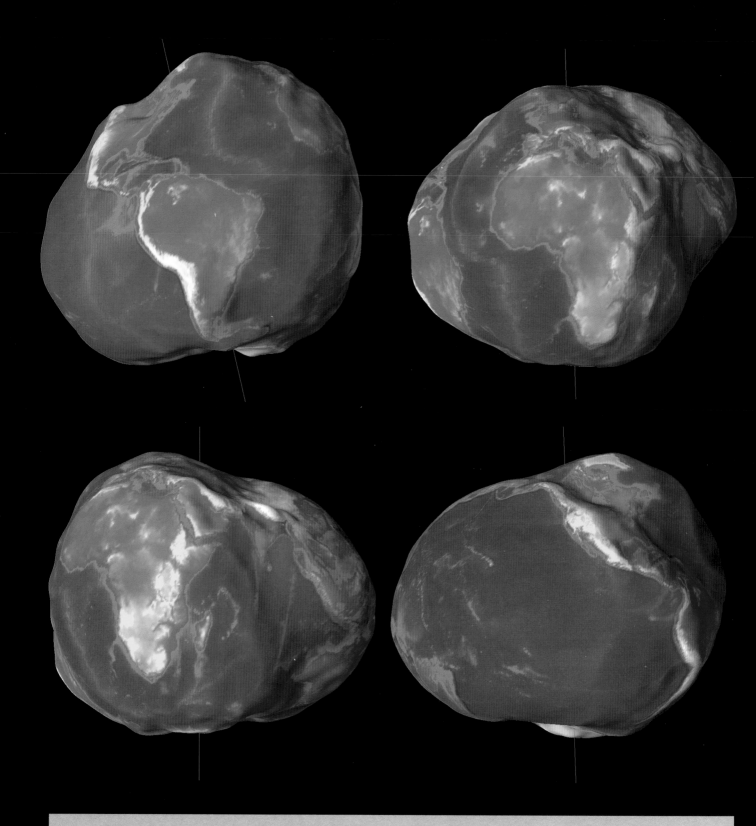

## THE SHAPE OF THE EARTH

These four computer-generated images show different views of the Earth, illustrating the shape (greatly exaggerated here) it would have if it were completely covered in water, referred to by geophysicists as the geoid – in the oceans, the geoid coincides with the present sea surface. The diameter of the Earth at the equator is slightly greater than the diameter between the poles, so the planet has an ellipsoidal shape. However, there are a number of depressions and bulges in this ellipsoid. When greatly exaggerated, these give the Earth a highly distorted shape. In reality, these dimples are not more than a few tens of metres deep.

## POST-GLACIAL REBOUND

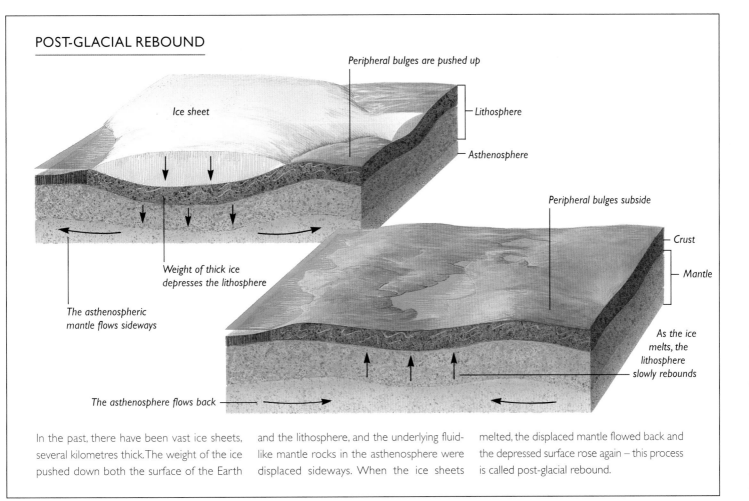

Ice sheet

Peripheral bulges are pushed up

Lithosphere

Asthenosphere

Peripheral bulges subside

Crust

Mantle

Weight of thick ice
depresses the lithosphere

The asthenospheric
mantle flows sideways

As the ice
melts, the
lithosphere
slowly rebounds

The asthenosphere flows back

In the past, there have been vast ice sheets, several kilometres thick. The weight of the ice pushed down both the surface of the Earth and the lithosphere, and the underlying fluid-like mantle rocks in the asthenosphere were displaced sideways. When the ice sheets melted, the displaced mantle flowed back and the depressed surface rose again – this process is called post-glacial rebound.

If one is prepared to accept that the interior of the Earth is capable of flowing like a fluid, then there is a simple explanation for these vertical movements: post-glacial rebound. This makes use of the fact that much of northern Europe and Canada was once covered by huge ice sheets, several kilometres thick, which started to shrink about 18,000 years ago. The weight of these ice sheets literally depressed the surface of the Earth, pushing down the bedrock and underlying mantle (there is also a peripheral zone where the surface has locally bulged up because the fluid mantle beneath the ice has been displaced sideways and upwards). This is rather like pushing your finger into toffee, creating a depression. When you remove your finger, the toffee starts to flow back, slowly filling up the depression. This is exactly what appears to be happening in Scandinavia and northern Canada. When the ice melted and dwindled away, the land surface slowly began to rise again as the fluid-like

rock in the Earth's interior started to flow back (see overleaf). The harbour tide gauges show that the depression created by the ice sheets is still rising today. In the last 7000 years, the centre of each depression has risen about 120 metres. It is estimated that these regions will have to rise roughly another 200 metres before the Earth has returned to its stable state.

So what about the evidence from earthquake vibrations? It turns out that there is no contradiction after all. If the Earth is shaken by an earthquake, over periods of seconds, the mantle rings like solid metal. When the huge force of gravity acts continually over geological time, the mantle rocks flow. An example of a more familiar substance that behaves in this way is 'silly putty' (silicon putty) which children play with. If you roll some of this into a ball and drop it on the ground, it will bounce. When it bounces it rapidly changes shape and vibrates, behaving just like the Earth when it 'rings'. However, if you leave the ball on

the ground, it will slowly flow away, becoming no more than a puddle of putty.

The notion that the Earth's mantle is capable of flowing like a fluid has turned out to be one of the key ideas in understanding the underlying driving force of the geological activity at the surface. The full implications of this fluid flow became clear when geophysicists started thinking about the temperatures inside the Earth.

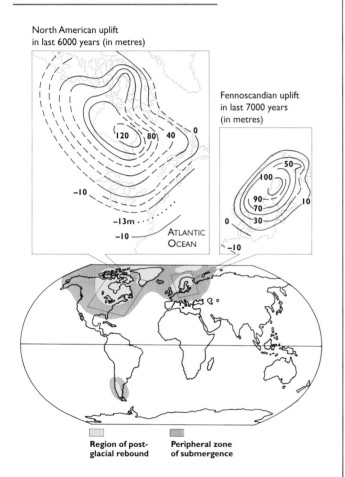

## POST-GLACIAL REBOUND IN NORTH AMERICA AND EUROPE

North American uplift
in last 6000 years (in metres)

Fennoscandian uplift
in last 7000 years
(in metres)

120  80  40  0

50

100

90
70  10

-10

0  30

-13m

ATLANTIC
OCEAN

-10

-10

Region of post-glacial rebound     Peripheral zone of submergence

Eighteen thousand years ago there were giant ice sheets in the northern hemisphere. The weight of the ice pushed down the Earth's surface, displacing the underlying mantle and creating large depressions surrounded by bulges. As the ice melted, the Earth's mantle flowed back, allowing the surface to return to its original shape. In the process old shorelines have been uplifted or submerged – the centres of the depressions have risen as much as 120 metres in the last 7000 years, while the bulges have subsided over 10 metres. These movements can be used to calculate the runniness (viscosity) of the Earth's mantle.

## THE INFERNO INSIDE

We began this chapter with two simple observations about the Earth. We now add a third.

*Third observation:* The deeper you go the hotter it gets. This is spectacularly demonstrated in the deep gold mines of South Africa, which penetrate up to 5 kilometres into the crust. Here, the temperature increases by about 10 to 15°C for every kilometre down, so that the deepest levels would be hotter than scalding water if they were not kept cool with ice-chilled air.

In Jules Verne's *Journey to the Centre of the Earth*, Professor Lidenbrock's nephew Axel points out that if the increase in temperature were to continue down-wards, the temperature near the centre of the Earth would be fantastically high. In fact, as we have seen, experiments show that the temperature near the centre of the Earth is only about 5000°C instead of the extrapolated 80,000°C (our calculation). The only explanation is that the increase of temperature must drop off at depth, so that the temperature gradient deep in the Earth is much less than near the surface.

There is a simple explanation for a decrease in the temperature gradient at depth in the Earth. To understand this, we need to recall our onion-like image of the planet, in which the solid mantle forms a shell between the crust and the molten outer core. One can think of the mantle as a vast tank of hot fluid which is being heated from underneath by the hot core, and cooled on top by the cold crust and atmosphere. Geophysicists have experimented with laboratory-scale versions of this. They have found that in certain situations the fluid may start to flow, so that hot material rises, cools and then sinks in a pattern of convection. The flow organizes itself into convection cells in which the upward rise of hotter fluid is balanced by a colder descending flow. Cells have central circular plumes of upwelling material surrounded by sheets of cooler downwelling fluid.

This also commonly happens in a saucepan of toffee or jam heated on a stove. Convection is a very effective way for a large body of fluid to lose heat and cool, and evens out any temperature gradients in the main body of the fluid. The temperature in a thin layer at the top will increase rapidly with depth, but from there on in the middle of the tank, the temperature will be more uniform. Thus, the temperature in a convecting Earth would also be expected to increase markedly with depth near the surface, but at deeper levels change much less.

Convection occurs because the density of the fluid is related to its temperature – the hotter parts of the fluid expand and become less dense, and the colder parts contract and are denser. A tank of fluid, heated from below, will be hotter and less dense near the bottom and colder and more dense near the top. Gravity will always tend to force the denser parts to sink, and the less dense parts to rise. However, any movement of the fluid is resisted by its stickiness. A very sticky fluid, like cold toffee, is just too stiff to convect. It turns out that there are a number of other factors which determine whether a fluid will convect, such as the temperature difference between the top and bottom of the tank, the depth of the tank, the force of gravity, and the thermal properties of the fluid (i.e. thermal conductivity and thermal expansion). These were first investigated by the physicist Lord Rayleigh near the turn of the century. He combined them into a single number, called the Rayleigh number. The importance of the depth of the fluid can be easily demonstrated by heating some cooking oil in an electric frying pan. If you tilt the frying pan, then you will see movement in the oil at the deep end, where it is convecting. But where the oil is too shallow, it will remain motionless. Lord Rayleigh showed that if the Rayleigh number for a tank of fluid is greater than about a thousand, the fluid will always convect.

If the Earth's mantle flows, then it should be possible to determine whether it is convecting by calculating its Rayleigh number. To do this, it is necessary to quantify all the relevant factors for the Earth. Some of these, such as the material properties and temperatures, can be measured in laboratory experiments on mantle materials. The size of the mantle is known from earthquake studies. But a crucial factor, which is difficult to measure directly, is the stickiness or viscosity of the mantle. The analysis of post-glacial rebound in Scandinavia and northern Canada, which we have described already, has proved to be the best way to estimate this. The rate at which

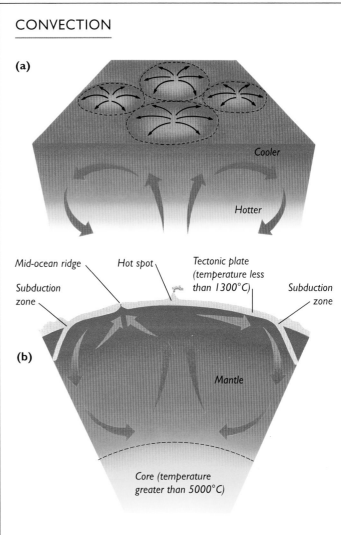

## CONVECTION

**(a)**

Cooler

Hotter

Mid-ocean ridge    Hot spot    Tectonic plate (temperature less than 1300°C)

Subduction zone    Subduction zone

**(b)**

Mantle

Core (temperature greater than 5000°C)

A tank of fluid, such as syrup, when heated from below, will often start to flow in a pattern of convection – this way the fluid carries heat from the bottom to the top of the tank. Hot plumes rise up from the bottom of the tank, then spread out sideways and cool before sinking back down again **(a)**. The Earth's mantle beneath the tectonic plates shows a similar pattern of convection – the rising parts form mantle plumes and upwellings beneath mid-ocean ridges, and downwelling parts coincide with the sinking plates in subduction zones **(b)**.

the land has risen is related to the viscosity of the underlying fluid-like mantle. To account for the observed uplift rate in regions several thousand kilometres across, the underlying mantle must have a viscosity of about 10,000 million million million poises (the poise is a unit of viscosity). This may not mean much unless you consider that water has a viscosity of about a hundredth of a poise. The Earth's mantle is a very, very stiff fluid. Despite this, the mantle is still large enough and hot enough to have a Rayleigh number which is abundantly supercritical – at least a thousand times greater than that needed for convection. In other words, Lord Rayleigh's theory demands that the mantle is not only convecting, but it is doing this vigorously, like a cauldron of hot tar!

## AN EXPLANATION FOR PLATE TECTONICS

A convecting interior to the Earth provides the fundamental driving force for plate tectonics. In fact, the motion of the plates looks itself like a form of convection. Hot material wells up at the mid-ocean ridges to create the ocean floor. The ocean floor then moves sideways and cools before sinking back into the Earth's interior at subduction zones. The plates exist because they are colder and stronger than the underlying mantle. This view of plate tectonics suggests that it is simply a mechanism by which the Earth slowly loses its heat.

Geophysicists can now catch the mantle, as it were, in the act of convecting. This is because, with the greatly expanded network of earthquake-monitoring stations and massive computing power, it is possible to use seismic vibrations to probe the Earth in virtually the same way as a PET scanner produces pictures of the human body. The image, called a tomographic map, is a three-dimensional picture of the variation in speed of body waves in the Earth. It is a rather blobby picture of the structure of the Earth, but this is an extra level of detail on top of the general onion structure determined from classic seismology.

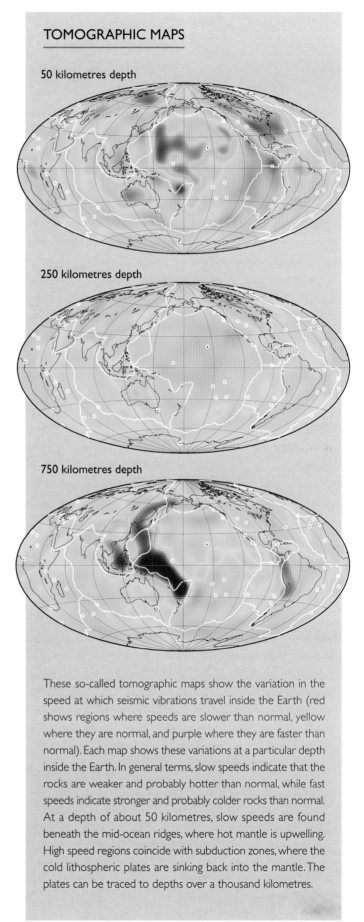

## TOMOGRAPHIC MAPS

**50 kilometres depth**

**250 kilometres depth**

**750 kilometres depth**

These so-called tomographic maps show the variation in the speed at which seismic vibrations travel inside the Earth (red shows regions where speeds are slower than normal, yellow where they are normal, and purple where they are faster than normal). Each map shows these variations at a particular depth inside the Earth. In general terms, slow speeds indicate that the rocks are weaker and probably hotter than normal, while fast speeds indicate stronger and probably colder rocks than normal. At a depth of about 50 kilometres, slow speeds are found beneath the mid-ocean ridges, where hot mantle is upwelling. High speed regions coincide with subduction zones, where the cold lithospheric plates are sinking back into the mantle. The plates can be traced to depths over a thousand kilometres.

The speed of body waves is determined by both the strength and density of the rock. But there is a relationship between temperature and strength, and so the small speed changes, seen in the tomographic maps, can often be thought of as a proxy for temperature variations in the mantle.

The tomographic maps give us an image of mantle convection which suggests that the flow in the deep mantle is not always tightly coupled to the movement of the surface plates. It is true that the sinking plates appear as cold tongues which extend down to depths of at least 1000 kilometres, forming the descending cold parts of giant convection cells. But the axes of the mid-ocean ridges do not coincide with the corresponding ascending hot parts. Instead, they seem to be associated with rather superficial upwellings confined to the upper mantle. The much deeper 'hot' patches, visible in the tomographic maps, do not form long walls of hot mantle that would match the linear shape of the mid-ocean ridges, but rather appear as isolated regions which do not connect up in an obvious way with these ridges.

It is possible that the internal layering within the mantle, especially the division between the upper and lower mantle where there is a change in the mineral structure, may play a part in breaking up simple mantle-wide convection cells. But even taking this into account, geophysicists find it difficult to get convecting fluids, modelled inside a computer, to spontaneously produce surface plates which look like the great tectonic plates on Earth. This requires an additional ingredient, not included in the computer models – something which lubricates the plates. For the tectonic plates to move relative to each other without breaking up, they have to slide past each other with ease. Also, strong plates, lubricated underneath, can move in a way which is partly disconnected from the underlying churnings of the deep mantle, sliding over a number of hot ascending plumes before sinking again in the cold descending limb of a convection cell.

The mantle beneath the plates is very close to its melting point. Indeed, it is likely that near the base of the plate, microscopic pockets of molten rock, sweated out of the hot mantle, are trapped. This

*A fountain of lava erupts from the volcano of Kilauea in Hawaii. The lava is runny molten basalt, forming a river of molten rock which flows down the flanks of the volcano. Repeated eruptions like this have built up the island.*

portion of the mantle, called the asthenosphere, would be weaker than normal, forming a sort of lubricated zone on which the plates can glide. Water may be a key factor in this. Its presence weakens rocks in both the mantle and in the crust, as well as promoting local melting. In the crust it penetrates the fractures where two plates slide past each other and acts like a sort of oil. In addition, the water reacts with the rock causing new minerals to crystallize on the fracture planes. These minerals are weaker than the surrounding rock, again behaving like a lubricant.

The role of water in plate tectonics begs an intriguing question. If water did not exist on the surface, would the Earth have the pattern of plate tectonics we see today? The answer is probably no. We will explore this conclusion further in Chapter 8, when talking about plate tectonics on other planets.

But what about the plumes of hot mantle rock that rise from the deep mantle, which seem to be another feature of convection apart from plate tectonics? Are they significant for anything we can see on the surface?

## MANTLE PLUMES

Geologists have long known about volcanoes, or volcanic chains, which seem to lie isolated in the middle of the great tectonic plates. For example, the great volcanoes of Kilauea and Mauna Loa lie on one of the Hawaiian islands in the middle of the Pacific plate. The volcanic rock which makes up the volca-

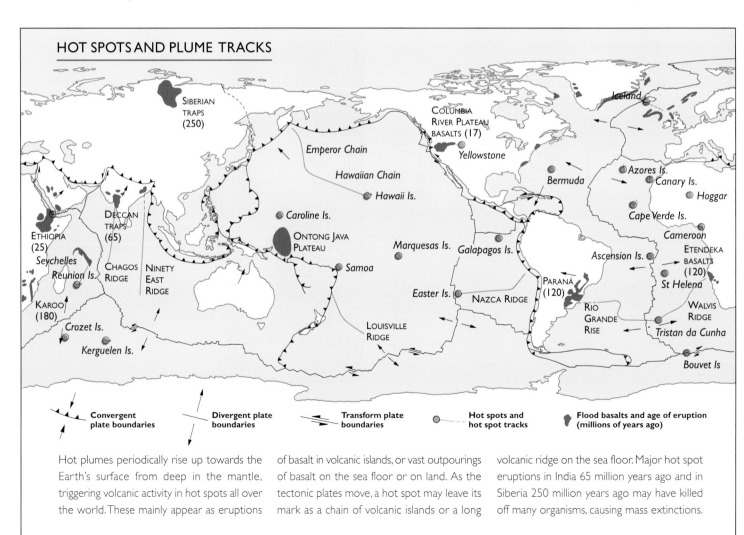

### HOT SPOTS AND PLUME TRACKS

Hot plumes periodically rise up towards the Earth's surface from deep in the mantle, triggering volcanic activity in hot spots all over the world. These mainly appear as eruptions of basalt in volcanic islands, or vast outpourings of basalt on the sea floor or on land. As the tectonic plates move, a hot spot may leave its mark as a chain of volcanic islands or a long volcanic ridge on the sea floor. Major hot spot eruptions in India 65 million years ago and in Siberia 250 million years ago may have killed off many organisms, causing mass extinctions.

*A river of runny basalt lava flows over cliffs of volcanic rock into the sea during a recent eruption of Kilauea on the island of Hawaii.*

noes is basalt – the rock produced when the Earth's mantle melts – prompting many geologists to propose that the volcanoes are the expression of some sort of hot spot in the underlying mantle. They are in fact the youngest of a line of volcanoes which runs for thousands of kilometres to the northwest. It seems that the hot spot has remained more or less fixed, acting almost like a blow torch, burning a line of volcanoes as the Pacific plate has glided over it. This suggests that the hot spot has been a long-lived feature of the mantle. What is more, a kink between the Hawaiian and Emperor volcanic chains seems to record a change in the motion of the Pacific plate. The volcanic island at the kink erupted about 45 million years ago, exactly when the drifting Pacific plate veered from a northerly to a more northwesterly course.

Geophysicists, with their new understanding of mantle convection, believe that the hot spot is the surface expression of a hot plume which has risen up from a much deeper level in the mantle. Evidence for these plumes has now been found all over the Earth. The plumes push up the surface into a broad dome, over 1000 kilometres across – the ocean floor in a wide region around the Hawaiian islands is about 1.5 kilometres shallower than the surrounding deep ocean. The volcanic activity is merely a pinprick in

*At the Iguazú Falls, near the border between Paraguay, Argentina and Brazil, the Paraná river flows over cliffs of basalt lava. This is just a small part of a wide region of lava which erupted about 120 million years ago above an upwelling plume of hot mantle as South America rifted from Africa.*

the centre of the plume swell. A suggestion that the plumes come from very deep down is provided by the helium gas which is released into the atmosphere during volcanic eruptions (see Chapter 3). As we have seen, helium is made up of two isotopes, helium-3 and helium-4. The proportion of these isotopes is a sort of fingerprint which can tell one about the original rocks which melted to feed the volcanoes. The helium released from the Hawaiian volcanoes is much richer in helium-3 than the helium emitted from

*Left: The volcanic island of Hawaii forms the youngest in a chain of volcanoes which have erupted over the last few tens of millions of years above a rising plume of hot mantle. This space shuttle image shows Hawaii in the distance, with some of the older volcanic islands in the foreground.*

the axis of the mid-ocean ridge. The source of the helium leaking from the crest of the mid-ocean ridge must be the upper mantle, where the tomographic maps suggest the hot upwelling beneath the ridges is confined. The difference between the plume and mid-ocean ridge helium suggests that the mantle, overall, is not well mixed, and there are distinct pockets that are tapped by the plumes. One idea is that the plumes come all the way from the bottom of the mantle. Here, at the interface between the solid mantle and liquid core, hot blobs of rock, rich in helium-3, may occasionally rise up to the surface.

The plume head must eventually melt to trigger the volcanic activity in Hawaii and other isolated volcanoes. How much molten rock it produces will depend on both the pressure and temperature –

## RIFTING, VOLCANISM AND MANTLE PLUMES

About 120 million years ago South America and Africa had already begun to split apart above an upwelling plume of hot mantle, triggering widespread volcanic eruptions **(a)**.

Subsequently, as the rifting plates moved over the centre of the plume, a meandering track of thick volcanic ridges formed. Today, the centre of the plume underlies the island of Tristan da Cunha **(b)**. The plume forms a wide mushroom-shaped body of upwelling hot mantle and the volcanic activity is confined today to the centre of the plume head, marking the hot spot on the surface **(c)**.

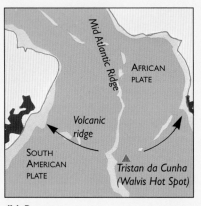

**(a)** 120 million years ago

**(b)** Present

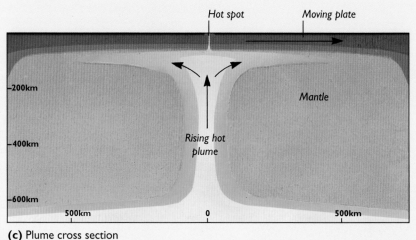

**(c)** Plume cross section

generally speaking, more melt will be produced when it is at a high temperature and low pressure. So melting starts when the plume reaches a level in the mantle where the pressure is low enough, given the temperature of the rock in the plume. Usually, the plume head rises no higher than the base of the lithosphere, at a depth of 100 kilometres or so. Here, the pressure is still high and only a small amount of melting occurs. If the plume head is exceptionally hot, the amount of molten rock will be much larger. But if the hot plume could reach the surface of the Earth, then the decompression it would undergo would be sufficient to cause melting on a very large scale. One place this could happen is near the crest of a mid-ocean ridge. For example, in Iceland, the mid-Atlantic ridge overlies a plume. The extra melting of the mantle here has been sufficient to build up the mid-ocean ridge above sea level, creating the island of Iceland over 2.5 kilometres higher than the summit of the mid-ocean ridge elsewhere.

If a hot plume came to the surface where two continents were splitting apart, there would be catastrophic volcanic activity. This seems to be exactly what has happened at various times in Earth history. For instance, a pile of basalt, over a kilometre thick, covers much of the western margin of India in the Deccan region. It is now known that these lavas erupted over a period of less than a million years, about 65 million years ago when India rifted away from the Seychelles Bank. Another spectacular example is found on the east coast of Greenland where lavas, extending for hundreds of kilometres, erupted about 60 million years ago when Greenland began to split away from Britain. In fact, the plume that caused this melting still exists; it is the one that underlies Iceland. There are many other examples throughout the world. On the eastern side of South America in Argentina, Paraguay and Brazil, the extensive lava flows of the Paraná region erupted when South America split away from southern Africa about 120 million years ago. The same volcanism is found in the Etendeka region of Angola on the other side of the South Atlantic Ocean, and the track of the plume that caused this can be seen in the ocean floor, creating the submerged Walvis Ridge.

The French geophysicist, Vincent Courtillot, has argued that without plumes, the continents would long ago have settled into a single landmass. When two continents collide, an even larger continent is created. Earth history has seen the formation of several supercontinents. But once formed, why should they break up? Courtillot believes that plumes provide the answer. When a plume approaches the surface, it causes the overlying plate to dome, creating, as we have already mentioned, a large bulge in the ocean floor. But continental plates are generally much weaker than the oceanic plates (see Chapter 5). Doming of a continent may be sufficient to cause the continent to literally fall apart, splitting the crust. The signature of these continental domes still exists. The rivers around the margins of the plume eruptions in India, South Africa and South America have a radial pattern as if they were flowing off a dome. Though the dome itself no longer exists, it seems that the rivers were established soon after the plume eruptions when the dome was still extant. In same cases, the uplift associated with the doming has become permanent, creating, for instance, the highlands of eastern South Africa. This may be because large quantities of molten rock never reached the surface, but remained frozen at the base of the continent. It has been calculated that this could be many times greater than the thickness of the lavas seen on the surface. In this case, the addition of this rock has permanently built up the continent – plumes may help to build up the continental crust.

Courtillot also believes that many of the great mass extinctions of life coincide with the volcanic eruptions triggered by these plumes. The eruption of the Deccan lavas coincides with the mass extinction at the end of the Cretaceous, when the dinosaurs among other creatures went extinct. It is tempting to link the two events. Certainly, the enormous amount of ash and fumes that would have been ejected into the atmosphere during the Deccan eruptions would have

had a global impact on the climate, perhaps darkening the sky, forming acid rain and killing off life. Courtillot has taken the idea even further, suggesting that the plumes actually drive the evolution of life (see Chapter 7), periodically causing mass extinctions and changing the configuration of the continents. The latter would open up new habitats and help life to achieve its present complexity and diversity. Whether or not these ideas are correct, plumes are certainly an important feature of the planet. In many respects, the shape of the Earth is like a giant golf ball, dimpled by the doming created by dozens of plumes.

We have now gone as far as we can in our long journey into the Earth's interior. We started out on the surface with some speculative ideas, but eventually found not just the driving force for plate tectonics, but a cause of giant volcanic eruptions, mass extinctions, continental break-up and growth. Our journey to the centre of the Earth ended up right back on the surface. But what did Professor Lidenbrock and his nephew Axel, in Jules Verne's adventure story, find on their journey to the centre of the Earth? They began their journey above a plume in Iceland, descending into the crater of the volcano called Sneffels. But their subsequent discoveries would confound all modern scientific expectation. The interior of the Earth is full of gigantic holes, partially filled with water. Here, the explorers discovered prehistoric monsters, left over from the time of the dinosaurs. After many adventures, Professor Lidenbrock and Axel eventually returned to the surface back up the crater of another volcano far away in Italy, thrown out during an eruption. The interior of the Earth envisaged by Jules Verne is an ancient and dank place, more like a network of underground caves. Geologists have now shown that it is a dynamic place, in constant motion. But Professor Lidenbrock would be undaunted by the discrepancy between his discoveries and modern science. After all, as he remarked himself: 'Science is eminently perfectible, and each new theory is soon disproved by a newer one.'

In the next chapter we will explore how the workings of the Earth's interior are linked to another prominent feature of the planet – huge mountain ranges in the continents.

### CHAPTER 5

# THE FLOW OF THE CONTINENTS

· · · · · · · · · · · · · · · · · · · · · · · · · ·

Most of the dry land on Earth sits no more than a
few hundred metres above sea level. But in some
places mountain belts rise to heights of several
kilometres. These regions are often prone to
devastating earth tremors. How are mountains
formed and what is the connection with
earthquakes? The answer may lie in the fluid-like
properties of the Earth's outer layers. According
to a new theory, mountains may flow up or down
when continents collide. In the process they affect
the circulation of the planet's atmosphere and
change the climate.

*The snow-capped peaks of the Southern Alps in New Zealand rise up several thousand metres*
*on the west coast of South Island. A major fault line, called the Alpine Fault, runs along the range front.*
*Today, the mountains are being both pushed up and dragged sideways along this fault.*

## A SHARP PAIR OF EYES

During the Great Depression of the 1930s, seventeen-year-old Harold Wellman was at a loose end. A recent English immigrant to New Zealand, he joined the many prospectors making for the wild west coast of South Island, where the Pacific Rim's latest gold rush was in full swing.

The gold was found on the many beaches down the coast, mixed up with sand. After every storm, the prospectors would go in search of a 'cut' – a new section of beach that had been excavated by the waves to reveal gold-rich black sand. Harold spent in all six months on the west coast, earning about a pound a day from his gold panning. Though the experience did not make him rich, it inspired in him a fascination with the natural world which has remained with him all his life. When his gold panning days finally ended, Harold Wellman acquired a degree in geology and joined the New Zealand Geological Survey.

During the Second World War, the Geological Survey sent Wellman back to the west coast of South Island to look for the mineral mica, which was in short supply. He travelled down the coast on foot. On a clear day, the high snow-capped peaks of the Southern Alps dominate the scenery, rising to nearly 4000 metres at the summit of Mount Cook. The steep range front forms an almost impassable barrier not only to travellers but also to the heavy rain clouds that are blown in from the Tasman Sea – the coastal region has one of the highest rainfalls on Earth, with up to 8 metres a year.

Harold Wellman's job was to map the different rock types that lay beneath the lush landscape. This was not an easy task because the bedrock was heavily disguised by the dense vegetation. But the numerous rivers which flow across the coastal plains from the Southern Alps to the sea were full of boulders. He noticed two types of boulder – schist and granite. Veins in the schist and granite were possible sources of the mica he had been commissioned to find. Schist

is a distinctive banded rock – sediment that has been deeply buried, compressed and heated up. The range front of the Southern Alps is almost exclusively made up of this rock type. The granite is a pale rock with a distinctive spotted appearance, made up of quartz and feldspar minerals with flecks of black mica. It was once molten. The granite boulders were coming from low conical hills at the foot of the Southern Alps. As Wellman started investigating the source of the granite more closely, he had an idea. He guessed that the boundary between the granite and schist was, in fact, a nearly straight line following the range front of the Southern Alps and extending for hundreds of kilometres all down the coast of this part of New Zealand (see opposite).

On his next trip, Wellman climbed up boulder-strewn creeks in search of the actual contact between the two rock types. This contact did, indeed, seem to follow his straight line. Wherever he got close to it, he found green clay in the banks of the creek which he could gouge out with his bare hands. In the north, a distinctive band of volcanic rock on the western side stopped abruptly at the line. Much further south, he found a very similar band of volcanic rock which also stopped abruptly at the line, but this time on the eastern side. These relationships suggested that his line was a fault, but this fault was much longer than any that had ever been discovered before, extending for nearly 700 kilometres. The granite and the schist had moved past each other, pulverizing the rock at the fault and creating the soft green clay. Wellman proposed that the distinctive bands of volcanic rock were once a single continuous band which had been displaced sideways 500 kilometres. He published his discovery in 1942, calling the new fault the Alpine Fault because it ran all along the foot of the Southern Alps.

Wellman realized that the motion on the Alpine Fault was not only sideways – it seemed that the mountains of the Southern Alps had also been pushed up along this fault line. Sideways and upwards displacements across fault lines had been seen elsewhere in New Zealand after great earthquakes. In 1855, a large earthquake damaged the fledgling city of Wellington at the southern tip of New Zealand's

North Island. In this earthquake, movement occurred along a major fault line to the east of Wellington, raising the coastline and creating a coastal platform, out of reach of storm waves, which cattle drovers could use to bring sheep to and from outlying farms. In 1922, a millrace in the northern part of South Island was broken by movement on a fault line during an earthquake. One side of the fault was uplifted over 3 metres and shifted sideways by about a metre so that it was no longer possible for water to flow along the millrace in the old direction.

The more Wellman looked at the landscape of New Zealand, the more evidence he found for fault lines and past earthquakes. His sharp eyes noticed features which had been ignored before: distinctive notches which ran along the sides of mountains, long low ridges which cut across river flats, abrupt bends in rivers, drowned forests and swampy ground. All these suggested some underlying force which was carving up the landscape. He began to develop a new field in geology – the study of recent Earth movements – and his field area was a landscape which was being literally cut up, displaced and raised along fault lines in the crust.

Virtually no part of New Zealand was fixed; the whole landscape was on the move. Wellman found a pattern in the movement. New Zealand was being sheared sideways and compressed, and in the process the mountains of the Southern Alps and their continuation in North Island were being pushed up. Wellman realized that if movements of around a metre can occur across fault lines during a single earthquake, then, given the frequency of earthquakes in New Zealand, simple arithmetic meant that these mountains could have risen above the waves in a very short period of geological time – perhaps only a few million years.

Wellman published his work on the active movements in New Zealand in 1955. At that time, his work was ignored because nobody could understand what was causing the movements he had observed. But in the 1960s, the theory of global plate tectonics was developed. Geologists now saw the surface of the Earth as a number of rigid plates which were moving relative to each other (see Chapters 2 and 3). New

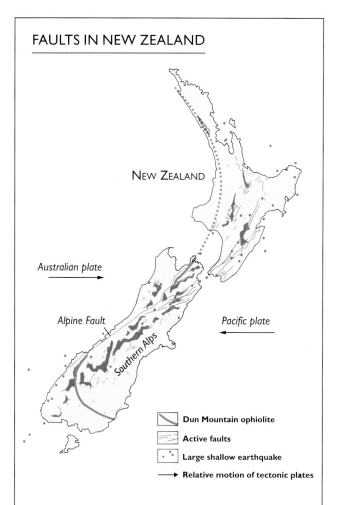

## FAULTS IN NEW ZEALAND

New Zealand lies on the boundary between two great tectonic plates – the Pacific and Australian plates. Here, the crust is broken up along fault lines which have ruptured during earthquakes, many in historic times. The Alpine Fault runs for several hundred kilometres as a nearly straight line along the western side of South Island. The Southern Alps are being pushed up along this fault, but the crust is also sliding sideways. A distinctive band of volcanic rocks in the northern part of South Island, called the Dun Mountain ophiolite, once joined up with a similar band much further south. It has been displaced nearly 500 kilometres by horizontal movement on the Alpine Fault. These movements are driven by the motion of the plates.

Zealand lies on the boundary between two great plates – the Australian and Pacific plates. The movements in the crust are driven by the motion of these two plates. In the oceans, the zones of moving crust between the great rigid plates are usually very narrow – less than 10 kilometres wide. But in New Zealand the movement zone was 250 kilometres wide, as wide as the islands themselves. At the time, this discrepancy was ignored by most geologists.

In 1965 Harold Wellman had a sabbatical year away from New Zealand and visited the western regions of Asia. With his wife, he travelled through the mountainous regions of Pakistan, Afghanistan and Iran. The years of work in New Zealand had trained him to read the landscape in a new way. This time he was also able to use aerial photographs taken during the last world war. There were clear signs of fault lines offsetting the landscape wherever he looked. Fault lines cut hills, displacing them sideways; fault lines deflected rivers and dammed lakes. In all, Wellman mapped a network of moving faults in a region 1500 kilometres wide. He published this work in 1966, and once again he was largely ignored.

Wellman's discoveries showed that profound changes to the landscape, such as the raising of great mountain ranges, could be accomplished relatively quickly, and that faults, and the earthquakes that occur along them, were the agents of this change. In the 1970s, armed with novel techniques which allowed them to study earthquakes without leaving their laboratories, a new generation of geologists began to appreciate the significance of Wellman's work.

## EARTHQUAKES EVERYWHERE

In the early 1960s, the partial nuclear test ban treaty between the United States and the Soviet Union placed limits on the size of the nuclear bombs both signatories could test. But given that they were exploded underground, the problem was to verify that the treaty was not being broken. A worldwide network of listening stations was set up by the US government to detect the vibrations from nuclear detonations. This became the WSSN (Worldwide Standardized Seismic Network) which routinely monitors all seismic

*Left: The Wairarapa Fault in North Island, New Zealand, forms a sharp ridge running diagonally across the picture from bottom left to top right – clear evidence of the mobile crust in New Zealand. The fault last moved in a great earthquake in 1855, displacing young river beds over 10 metres sideways.*

vibrations. Measurements at stations throughout the world are fed to a central recording agency in the United States where they are analysed. When the network was first set up it was immediately swamped by the vibrations from a myriad of earthquakes all over the Earth.

It was crucial to both locate the source of the seismic vibration and distinguish between a natural earthquake and a nuclear explosion. The location of the earthquake can be accurately determined by comparing the time when seismic vibrations first reach seismometers spread out over the world. But to distinguish a nuclear explosion from a natural earthquake requires careful analysis of the vibrations. A seismometer essentially consists of a free arm which can move in one of two directions – movement in one direction may indicate a fleeting compression of the underlying rock; movement in the other, momentary extension. In a nuclear blast, a pressure wave is created in the surrounding rocks which spreads out in all directions simultaneously from the source of the explosion. This wave travels through the Earth and is detected by the sensitive seismometers in the WSSN. When the wave front first reaches a seismometer, it causes the free arm of the seismometer to be deflected in a certain direction, referred to as the 'first' motion of the seismic wave. Because the explosion triggers a concentric outward compression in the rocks – this is the P wave we have already described in Chapter 4 – if all the seismometers are identical and wired up in the same way, they will all be deflected in the same way when the wave front first arrives and hence have the same first motion. This is the significance of the 'standardized' aspect of the WSSN. It allows detectors from all over the world to be used together to study seismic vibrations.

In almost all natural earthquakes, the vibrations in the surrounding rocks are triggered by sudden movement on a fault. In this case, the pattern of P waves which radiate out from the earthquake is different from that in a nuclear explosion. Because there is no uniform outward 'push' triggered during fault movement, in some directions from the epicentre the first distortion of the rock will be a compression, in others it will be an extension of the

## EXPLOSIONS AND EARTHQUAKES

**(a)** Explosion

**(b)** Earthquake during fault rupture

A nuclear explosion causes an initial outward compression of the rocks in all directions, radiating out from the source of the explosion as a wave (P wave). An earthquake is triggered by sudden movement on a fault in the Earth's crust. P waves radiate out from the source of the earthquake, but, unlike an explosion, they start moving in some directions with a brief extension of the rocks, and in others with a momentary compression. Seismologists can determine the direction of movement along the fault by analysing this pattern of waves.

rocks. Seismometers which lie in certain parts of the world will show one first motion and those in the other parts will show the opposite first motion. The precise position of a seismometer with a particular first motion can be used to determine the type of fault that moved in the earthquake. Geologists recognize three types of fault (see pp. 122–3). The crust is pulled apart along normal faults – the two sides of the fault move away from each other, often creating rifts in the Earth's surface. It is pushed together along reverse faults – this movement builds mountains. Finally, the crust slides horizontally along strike-slip faults – the Alpine Fault in New Zealand, discovered by Harold Wellman, has moved in this way.

The scientific benefits of the WSSN were enormous. For the first time, it was possible to accurately determine precisely where the earthquake-prone parts of the Earth were. Narrow bands of shallow earthquakes were detected in the oceans, following both the mid-ocean ridges and their offsetting transform faults. But wide zones of earthquakes were detected in all the continents. For instance, a belt of earthquakes extends in a great swathe through the mountainous regions of Morocco and Alpine Europe, on into Greece, Turkey, Iran and the mountain ranges of central and southeast Asia. Another belt follows the mountainous western margin of the Americas. In simple terms, where there are large mountain ranges, there are earthquakes. Each earthquake is the result of movement on a fault. The wide zones of moving faults that Harold Wellman had painstakingly found in New Zealand and parts of Asia are in fact just small parts of the global zones of earthquakes.

The pattern of earthquakes gives us an image of the continents which is not exactly what we would expect from the theory of plate tectonics. Though in very general terms we can identify rigid plates in parts of the continents, such as northern Europe and Africa, the eastern parts of both North and South America and southern India, substantial parts of the continents are not rigid at all. It is clear that if we are going to explain what is going on here we need a new theory for the continents. And this theory will be the key to understanding the origin of mountains.

## MOUNTAINS THAT FLOW

In the 1970s, geophysicists such as Dan McKenzie and Peter Molnar started thinking about the significance of the earthquakes in the continents. At that time, the theory of plate tectonics was still new. They tried to explain the earthquakes in terms of relative movement between many very small rigid plates – microplates – which appeared to be jostling together. But the earthquakes were so close together that they needed microplates only a few tens of kilometres across. These were tiny compared to plates in the oceans, which were thousands of kilometres across.

By the early 1980s, some geophysicists had started experimenting with a different way of thinking about the continents. At this time, Dan McKenzie was collaborating at Cambridge University with a young geophysicist, Philip England. They decided that the

### EARTHQUAKES IN THE CONTINENTS

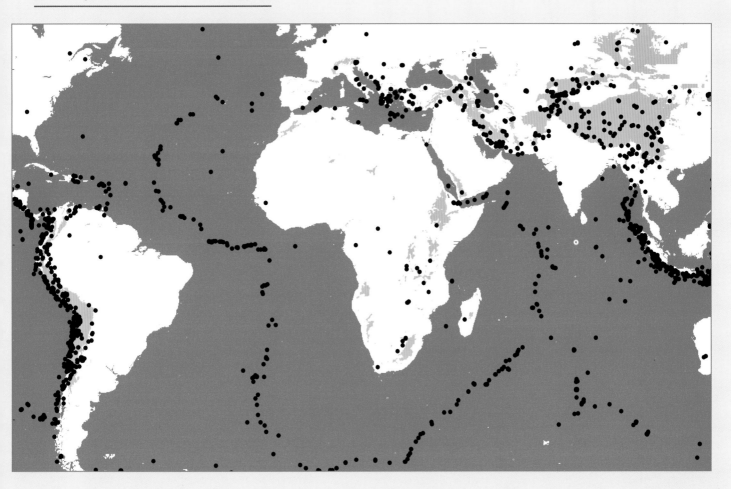

During the last thirty years, earthquakes throughout the world (shown by black dots in this computer-generated image) have been routinely detected. In the oceans, they occur in narrow zones which follow the mid-ocean ridges. However, there are wide earthquake-prone regions in the continents – these coincide with mountainous regions which are on average over 2000 metres above sea level (shaded yellow). For example, the Alpine–Himalayan chain of mountains, extending from the Mediterranean into Asia, is an extensive belt of earthquakes up to several thousand kilometres wide. Another wide zone of earthquakes follows the western margin of the Americas, especially along the Andes.

movement in the continents was so pervasive that perhaps it can be treated as a pattern of flow. Perhaps the continents behave more like a fluid than a rigid plate? At first, this may seem absurd. Everyone knows that the ground beneath their feet is rock solid. But often solid materials can exhibit fluid-like behaviour. It is really a question of the level of detail that you choose to look at the material. Take an everyday substance like sugar or flour. Sugar consists of many solid grains, but you can pour it out of the bag. In other words, the bulk behaviour at a large scale is fluid-like, even if at the small scale the individual grains are solid.

Now, everything on the Earth's surface is subject to the force of gravity – objects will fall if left un-supported. Thus any fluid on Earth is always pulled by the force of gravity and, unless it is pushed up, it will tend to flow away under the influence of this force. The fluid will stop flowing only when it has a level surface, like the surface of a lake or the sea. For example, imagine filling a tank full of honey. When you start pouring, the honey may form a low mound. But this will not last long because the honey flows out to fill the tank and will only come to rest when it has a flat surface.

The surface of a continent with mountains is clearly not level. So, if it is a fluid, it must be flowing, though it may be very sticky or stiff (i.e. have a high viscosity) – far stiffer than any fluid we are likely to encounter in everyday life. Thus if we accept the notion that continental crust can behave like a fluid, we must move away from the idea of mountains as static and fixed. They become features of the Earth's surface which exist while the continents are flowing. This, of course, is exactly what we observe. In New Zealand, or the Alpine–Himalayan belt of mountains, or the high Andes of South America, the mountains coincide with the zones of crustal earthquakes and young fault lines. And the earthquakes and young fault lines, as Harold Wellman first discovered over forty years ago, tell us that the landscape is moving.

## Faults in the landscape

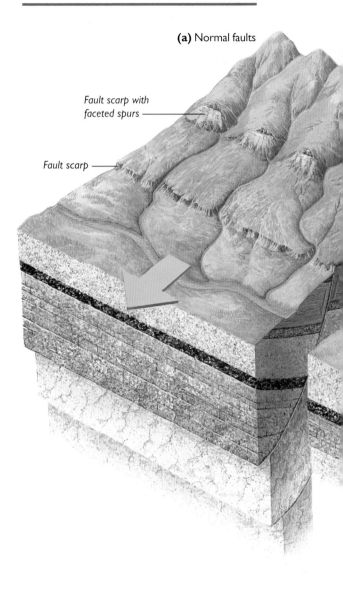

**(a)** Normal faults

Fault scarp with faceted spurs

Fault scarp

The Earth's crust is broken along faults. The faults displace not only the rock strata beneath the landscape, but the landscape itself. The crust is stretched by movement along faults called normal faults **(a)**. Movement on these faults creates depressions bordered by ranges of hills with sharp edges. The crust is pushed together along reverse faults **(b)**. Movement on these faults pushes up linear ranges of hills, locally blocking or deflecting the flow of rivers. The crust slides sideways along strike-slip faults **(c)**. Many features of the landscape, such as ridges, old river banks and even fields and fences can be offset by this sideways motion. Locally small lakes, known as sag ponds, may form in depressions along the fault line.

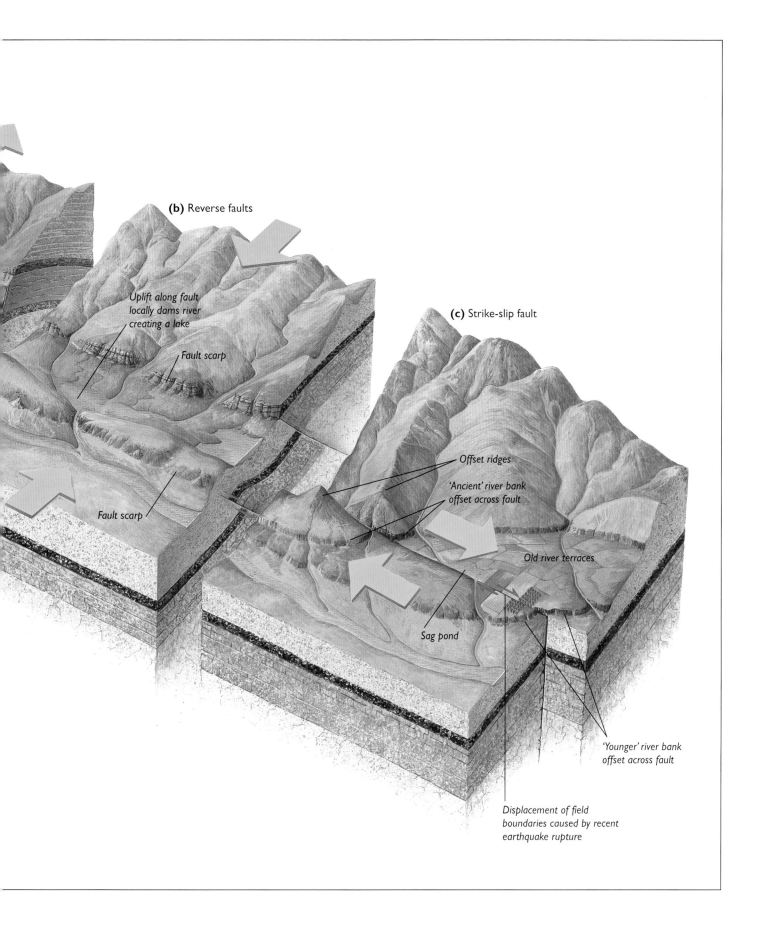

**(b)** Reverse faults

Uplift along fault
locally dams river
creating a lake

Fault scarp

Fault scarp

**(c)** Strike-slip fault

Offset ridges

'Ancient' river bank
offset across fault

Old river terraces

Sag pond

'Younger' river bank
offset across fault

Displacement of field
boundaries caused by recent
earthquake rupture

## JOURNEY ACROSS
## THE ROOF OF THE WORLD

An important test of the fluid theory is to see if it can work for the largest mountain ranges on Earth, those in central Asia. You have to journey right through them if you really want to appreciate their overall size and shape. If you travel north across the Ganges plain of northern India, you will traverse over 200 kilometres of flat country at an elevation of a few hundred metres above sea level, criss-crossed by numerous rivers. Eventually, you will see a long low ridge in front of you, rising up out of the plains. This ridge is the most southerly tip of the Himalayas. Behind it lies the mountain kingdom of Nepal and range after range of mountains, each one slightly higher than the one before. In the far distance, you will see the snow-capped 8000-metre peaks of the high Himalayas, apparently floating in space. You are looking at the highest mountain front in the world. To appreciate the height of these mountains, just look out of the window of a modern jet when it is at its cruising altitude. The ground seems very far below, and yet you are only slightly higher than the summit of Mount Everest.

The foot of the frontal ridge, marking the northern margin of the Ganges plain, is a fault line which runs in a great arc right along the southern edge of the Himalayas. At intervals along this fault line, rivers flowing from the Himalayas have cut deep gorges. Jean-Phillippe Avouac, a French geophysicist, has been studying these rivers. He has found that high up on the sides of the gorges, ten of metres above the present river, are the remains of old river beds. However, these beds no longer slope towards the Ganges but are warped into great arches. Pieces of wood in the river beds, dated using the carbon-14 method, are only a few thousand years old. Avouac has suggested that the warping and uplift occurred during earthquakes when there was movement along the great fault line at the foot of the frontal ridge. He estimates that the Himalayas are advancing over the Ganges plain at about 2 centimetres a year along this fault line. This is slightly less than half the total movement between India and Siberia predicted by the theory of plate tectonics. The remaining motion must be taken up in the regions further north.

North of the frontal ridge, the foothills of the Himalayas are heavily forested with subtropical vegetation. In the kingdom of Nepal, the steep hills are intensely cultivated, covered in a staircase of man-made terraces. The annual rainfall here is several metres, and every year, during the Indian monsoon, the landscape is deluged by heavy rain which cascades off the hills and turns the rivers into raging torrents. These rivers flow along valleys parallel to the mountain front before cutting across the foothills, carrying huge loads of silt to the Indian plains.

North of Kathmandu, the capital of Nepal, if you follow any of the big rivers upstream, you will eventually enter a deep gorge between the barren peaks of the high Himalayas. Here, the rock is made up of schist or granite, or folded and faulted marine sediments which were deposited over 60 million years ago on the margins of a vast ocean – the Tethys Ocean – that once lay between India and Asia. Their sculpted remains form the great peaks of Annapurna, Manaslu, Shishapagma, Cho Oyu, Everest, Lhotse and Makalu, which are all over 8000 metres high. The granite is the solidified remains of crust which was molten at depths of several tens of kilometres about 20 million years ago. It has been pushed up along giant faults which plunge under the high Himalayas. The river gorges cut deep into the rock, revealing the layering which is also inclined to the north. The sides of the gorges are thickly vegetated. But you will notice that high up above you, at altitudes of 3000 metres or more, the vegetation thins out, and higher up still the hillsides are scrub and eventually just scree and bare rock. This change in the vegetation marks the level at which conditions are too cold for most plants.

As you trace the river even further upstream, it becomes narrower with less water. And when you have crossed the imaginary line which links all the high peaks, the rivers are reduced to a trickle and the landscape becomes barren with very little vegetation. North of here, you enter a high desert of bare rocky

*The snow-covered peaks of the high Himalayas, here shown in Zanskar, rise locally over 8000 metres above sea level. Intervening glaciers have carved deep valleys. The mountains have been pushed up as India has ploughed northwards into Asia.*

View of the Himalayas and the great plateau of Tibet, looking west from the space shuttle. Heavy rain falls on the Himalayas, but the Tibetan plateau further north is a high desert, though lakes mark elongate depressions where the Tibetan crust has rifted apart. The general east–west 'grain' of the crust, where rocks have been pushed up along older faults, is also clearly visible in the plateau.

hills with occasional lakes. This is the beginning of the high plateau of Tibet which extends for thousands of kilometres to the north. You have also crossed a great topographic barrier which dams moist air blown from the Indian Ocean. The clouds build up against the range front of the high Himalayas, releasing their cargo of water. Further north, it rarely rains and the rivers are fed mainly by melt water from high mountain glaciers. Here and even further north, in central and northern Tibet, where parts of the plateau

form huge tracts of rolling countryside, the average elevation is over 5000 metres.

Travelling is difficult for visitors to Tibet as there are very few roads, virtually no petrol stations and no railways or airports outside Lhasa. All expeditions start in the old capital where the Dalai Lama used to live. You must buy all your food and petrol here if you want to travel to the outlying regions. It is usual to go in a convoy of two or three jeeps and a truck. The truck carries all the baggage and supplies, and you

Tibet is a high desert, generally over 5 kilometres above sea level. Elongate lakes lie in depressions where the crust is splitting apart in an east–west direction.

must be prepared to travel for thousands of kilometres without coming near any sign of permanent habitation. The roads are bad, often no more than rutted sandy tracks which shift from year to year, looking like loosely braided hair laid out across the landscape. The air is so thin that it is difficult to sleep at night. But you have the sense of being in one of the last great wildernesses on Earth. At night the only sound is the wind; the stars are unbelievably bright, and occasionally a shooting star traverses overhead.

If you drive north out of Tibet along the old trading route to Kashgar, you pass through the dry Kunlun mountains that border northern Tibet. From there on, you cross another major fault line in the crust – the Altyn Tagh fault – and leave the high plateau, descending into the great depression of the Tarim desert at an altitude of about 1500 metres. Still further north are the Tien Shan mountains, which form a last barrier between you and the great lowland plains of Mongolia and Siberia. If you complete the whole trip from northern India to Mongolia, you have crossed the widest and highest mountain range on Earth.

## MAKING TIBET OUT OF A FLUID

You would be forgiven for thinking, after completing the journey described above, that the massive ranges of Asia do not look like a fluid. But Philip England was one of a group of geophysicists who wanted to show that they do. He realized that the topography of this region is simpler than it might seem at first. He was not interested in the detailed shape of individual mountain peaks, rather the smoothed-out form of the whole mountain range. This is like the shape you would get if you could remodel the region with a giant bulldozer, pushing all the mountain tops into the valleys and sculpting a general bulge in the Earth's surface. In this view of central Asia, the most prominent feature is the Tibetan plateau, which is a remarkably flat region, nearly 2000 kilometres wide and at an average elevation of slightly over 5 kilometres.

It should always be possible to calculate the way any fluid will flow, and the shape of its surface, if one knows both how runny it is and the forces which drive the flow. This is what Philip England tried to do for central Asia, using the known behaviour of fluids and Newton's laws of motion, recreated inside a computer. He wanted to see if the fluid could be pushed up to form a mound which looked like the smoothed-out shape of the Himalayas and Tibet. He used the theory of plate tectonics as a starting point for this test. Studies of the age of the ocean crust in the Indian Ocean show that 84 million years ago India lay far to the south of what is today Asia, and the Tethys Ocean existed between these two continental masses. Subsequently, India advanced northwards towards Asia at about 10 centimetres a year, and the Tethys Ocean began to close up. About 55 million years ago India made contact with the southern margin of Asia and slowed right down, but continued to move northward at about 5 centimetres a year. The remains of marine sediments deposited in the old Tethys Ocean were squeezed up and are presently found, folded and faulted, in the high Himalyas and southern part of the Tibetan plateau. The age of the youngest marine sediments dates the last moments of the Tethys Ocean in this region – subsequently it ceased to exist.

Over the last 55 million years, India seems to have behaved like a rigid piston, pushing its way into Asia. We can quickly see how we would expect a fluid-like Asia to behave in this situation by imagining the effects of pushing a paddle through sticky honey. Advancing India is the paddle and the rest of Asia is the honey. The honey will build up in front of the paddle and, as we continue to push, the mound of honey will get wider and higher, creating a mini mountain range. The height of the fluid mound is determined by the viscosity of the fluid – the stiffer the fluid, the higher the mound. The force of gravity also controls the height. For instance, on the Moon, where the force of gravity is less than on Earth, we would expect the fluid mountains to be higher. Finally, the speed of the paddle is important – the faster the paddle moves through the fluid, the higher the mountains. The moment we stop pushing, the pile of honey will flow away leaving a flat surface in front of the paddle.

Here, we need to know something else about mountains. Remember that the outer part of the Earth is composed of a crust overlying the mantle. Geologists can tell the thickness of the crust by studying earthquakes, in the same way that they have probed the Earth's deep interior (see Chapter 4). The continental crust is always thicker beneath high mountains. For example, the crust beneath the high plateau of Tibet is about 70 kilometres thick compared to 35 kilometres beneath the plains of India. As a rule of thumb, if we take the thickness of the Indian crust as standard, then the extra thickness beneath mountain ranges is seven times the general elevation of the mountains. There is a good reason for this. The crust is less dense (with a density on average about 2.8 times that of water) than the underlying mantle (which has a density of about 3.3 times that of water) and more or less floats on top. But, like a floating iceberg in water, the amount that pokes up above the general land level – or sea level in the case of an iceberg – is much less than the underlying root. The *Titanic* discovered this general principle, with disastrous consequences, when it hit an iceberg and was holed by the huge underwater icy root. In some ways, the shape of the crustal root is like the reflected, but exaggerated, image of a mountain range in a still lake.

The presence of roots beneath mountains has an important implication for fluid-like continents. If mountains are to grow, they must flow in such a way that the thickness of the crust increases. So, as the mountains are pushed up, rock deeper in the crust is also pushed down into hotter regions of the Earth's interior. Eventually this rock becomes hot enough to melt – this way many of the granites found today in the high Himalayas may have formed. Flow which decreases the thickness of the crust will create the opposite of mountain ranges – depressions in the Earth's surface. These principles are embodied in Philip England's analysis. He found that a fluid which has similar properties to honey, but is much much more viscous, will pile up in front of the advancing paddle of India, looking a bit like the high ranges of Tibet. But in detail it produces a fluid pile that would be too wide, if scaled up to the size of Asia, and which slopes away from the edge of the paddle without forming a plateau.

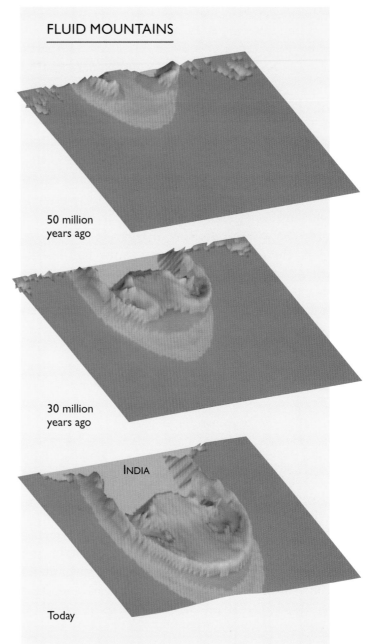

**FLUID MOUNTAINS**

50 million years ago

30 million years ago

INDIA

Today

Computer models of the formation of mountains successfully predict the main features of the great ranges of Asia. This model treats Asia as a fluid which is being pushed up by the advancing rigid block of India (shown by light blue region). The images illustrate a series of steps in the evolution of central Asia over the last few tens of millions of years, colour coded for height from highest (white) to medium (brown) to lowest (green). In the early stages, fluid-like Asia formed a narrow and generally low mountain range in front of advancing India. Later the mountains began to reach a maximum height, and from then on the ranges got wider, rather than higher, forming the great high plateau of Tibet.

Philip England had to resort to fluids with less familiar flow properties in his attempts to simulate in the computer the broad, high and flat surface of the Tibetan plateau. These fluids have the peculiar property that the harder you push, the weaker they become – there are certain paints which behave like this. The effect of this is to concentrate the fluid flow into a more focused region in front of the advancing paddle.

The computer models, using these more unusual fluids, outlined a history for the growth of the great ranges of central Asia. As rigid India moved northward into fluid-like Asia, a range of mountains grew in front of the advancing 'piston'. At first the mountains were relatively low and narrow. But as the movements continued, they progressively increased in height until they reached the maximum height that can be supported by fluid-like Asia. After that, as India carried on advancing, the existing mountains, instead of getting higher, became progressively wider. In this way, the Tibetan plateau was built up. But, to understand why Asia should behave as such an unusual fluid, it is necessary to examine in more detail the material with which fluid-like Asia is made.

## A SLAB OF TOFFEE

An important component of the rocks in the continental crust is quartz. This is the same mineral which forms the grains of sand on a beach – a beach is a good place to find fragments of the continental crust. So, if we want to understand how continents flow, we need to know something about quartz. Quartz at low temperatures will break along fractures when squeezed in a hydraulic press. This is very much what is happening in the crust during earthquakes. But the temperature in the crust rises progressively with depth, and at higher temperatures quartz changes its behaviour and can actually flow. This fits in with the observation that earthquakes are mostly confined to the top part of the continental crust, in a layer only a few tens of kilometres thick. The base of the earthquake-prone layer must coincide with the depth at which quartz is so hot that it starts to flow – this temperature is estimated to be about 350°C. Beneath the continental crust, in the mantle, the Earth is dominated by the mineral olivine (see Chapter 4). Laboratory experiments show that this mineral also flows at the high temperatures that exist in the mantle, but is a much stickier fluid than quartz. At lower temperatures, olivine is also stronger than quartz.

So the deep parts of the continents really are fluid. This flow is something that happens in the solid state – the quartz or olivine crystals change shape but preserve their special ordering of atoms. An example of a common material that does this is lead. Lead at room temperature is a solid metal, made up of a regular arrangement of atoms. Yet it can be bent or stretched – this becomes even easier if you heat the lead. When you bend lead, you are actually making it flow in the solid state. The ease with which quartz or olivine can flow is markedly increased if microscopic droplets of water penetrate the crystals. Because water is so abundant near the surface of the Earth, almost all natural quartz is weakened in this way. This too makes it likely to flow.

It seems as if the continents are rather like a slab of soft toffee. What we are really talking about is the portion of the Earth, called the lithosphere, which is usually strong enough to form rigid plates (see Chapter 4), though this is clearly not always the case in the continents. The bulk of the continental lithosphere is capable of flowing like toffee, but there is a thin covering layer, where the brittle crust breaks up along faults, which is more like a sprinkling of nuts on top. The whole slab is over 100 kilometres thick. With this image in mind, it is clear that the flow of the continents should be controlled by the fluid properties of the toffee portion. The brittle crust, like the nuts on the toffee, merely rides on top. Olivine has the fluid flow properties needed to model the shape of the Tibetan plateau – it becomes weaker the harder you push.

We are now in a position to say why the lithosphere beneath the continents can be fluid-like, while in the oceans it forms rigid plates. This is because the continents have a thicker crust than the oceans; generally over 35 kilometres compared to about 7 kilometres in the oceans. In simple terms, a substantial thickness of the continental plate contains weaker minerals such as quartz, and underneath this there is olivine at high temperatures – together these are weak enough for the whole lithosphere to flow like a fluid when caught between two moving stronger plates. Instead of weak quartz, the top part of the oceanic plate is mainly olivine at much shallower depths and lower temperatures than that in the continental plate – this cold olivine is strong enough to maintain the rigidity of the plate. This is not the whole story because, as we have seen, most of the Indian subcontinent has also acted as a rigid piston, pushing its way into Asia. The explanation here seems to be that India is a very ancient continent and has been around for thousands of millions of years. During much of this time, it has cooled and, because strength and temperature are closely related, become stronger than much of Asia.

The presence of thick continental crust has one other consequence: it makes the continental lithosphere less dense overall than the underlying deeper parts of the mantle in the asthenosphere. This is because the minerals in the continental crust are generally less dense overall than the olivine in the mantle. Therefore, unlike the lithosphere beneath the deep oceans, continental lithosphere cannot sink back into the Earth's interior and be subducted (see Chapter 3). So when two continents collide, the lithosphere is squeezed and thickened, pushing up large mountains.

## A SURVEYOR'S NIGHTMARE

Harold Wellman, that pioneer of the study of the moving crust, developed a technique in New Zealand which could provide a further test of the fluid theory. Before Wellman became a geologist he trained as a surveyor. In fact, much of his unusual approach to geology probably came from this training. In the early 1950s, he heard that a surveyor had been resurveying an area which had originally been surveyed some

*Scientists today can directly measure the movement of the Earth's crust. In New Zealand, satellite receivers (left of picture) have been used to accurately find the position of triangulation points on the tops of mountains – these points have changed position over time, moving a few centimetres a year.*

forty years earlier. He had found systematic differences between the two surveys, which he put down to inaccuracies in the first one. Harold Wellman had the idea that the differences might be due to movements in the crust.

New Zealand was in the unusual position of having very accurate early surveys, made by the British Survey of India at the end of the last century. The Victorian surveyors were determined to fix the landmarks of the young colony with as much precision as was possible at the time. The surveyor's technique was to climb to the top of the highest hills and sight from one hill to another. He can measure the horizontal angle between two lines of sight. In this way, the surveyor builds up a survey of the angles in a network of triangles, defined by the lines of sight between many hill tops. He can check his accuracy by adding up the angles in any triangle; they should add up to 180 degrees. The early surveyors could measure angles with remarkable precision. If there has been movement of the Earth's crust in the survey area, then the angles in the triangles will change. Harold Wellman realized that if the time between repeat surveys was

# The flow of the continents

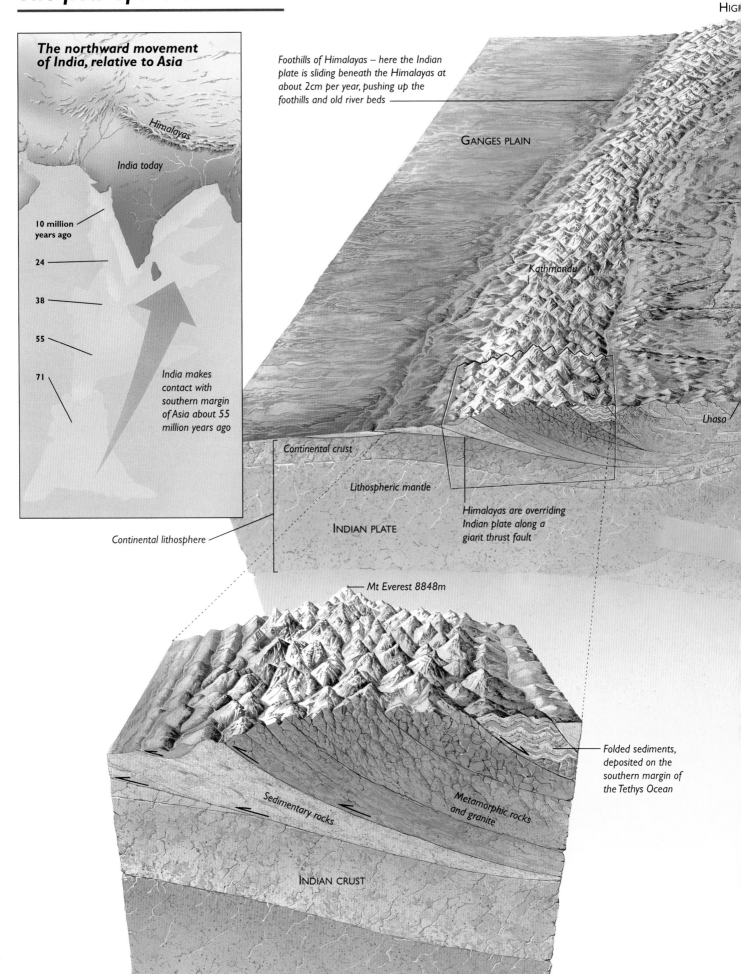

## The northward movement of India, relative to Asia

Himalayas

India today

10 million years ago

24

38

55

71

India makes contact with southern margin of Asia about 55 million years ago

Foothills of Himalayas – here the Indian plate is sliding beneath the Himalayas at about 2cm per year, pushing up the foothills and old river beds

GANGES PLAIN

Kathmandu

Lhasa

Continental crust

Lithospheric mantle

INDIAN PLATE

Himalayas are overriding Indian plate along a giant thrust fault

Continental lithosphere

Mt Everest 8848m

Folded sediments, deposited on the southern margin of the Tethys Ocean

Sedimentary rocks

Metamorphic rocks and granite

INDIAN CRUST

HIMALAYAS    KARAKORAM RANGE    TIEN SHAN MOUNTAINS

*To Kashgar*

TAKLAMAKAN DESERT

TARIM BASIN

*Altyn Tagh strike-slip fault
follows margin of Tarim basin*

PLATEAU OF TIBET

KUNLUN MOUNTAINS

ALTUN MOUNTAINS

*North–south rifts
where Tibetan plateau
is splitting apart in an
east–west direction*

*Suture between Indian
and Asian plates where
the Tethys Ocean
finally closed up*

*Region of faulting where
Asian plate is sliding
beneath Qilian
Mountains*

QAIDAM BASIN

QILIAN
MOUNTAINS

*Numerous strike-slip
faults cut Tibetan crust*

*Continental crust*

*Lithospheric mantle*

*Continental
lithosphere*

*Region where lithospheric root has detached
itself, sinking into the underlying asthenosphere*

ASIAN PLATE

*Asthenospheric mantle*

## Raising the great ranges of Central Asia

The great ranges of Central Asia have been pushed up as India has
moved northwards into Asia. India first made contact with the
southern margin of Asia about 55 million years ago. Since then it
has advanced a further 2000 kilometres, squeezing and thickening
the continental crust in a region which is now several thousand
kilometres wide, pushing up the mountains. Overall these
movements form a pattern of flow, but, in detail, they have taken
place in the crust by motion on numerous faults.

Today the plains of India are sliding underneath the Himalayas
along a gigantic fault line which runs along the southern edge of
the Himalayas. The fault slopes gently down towards the north
beneath the high Himalayas. Metamorphic rocks and granite from
deep in the crust have been pushed up along many other faults,
as the Himalayas and Tibetan plateau have risen to their great
elevations over 5 kilometres above sea level.

The crust beneath the Tibetan plateau is also splitting apart
along north–south rifts. Here the lithosphere may be thinner than
normal because the bottom part has detached itself, sinking into
the underlying asthenosphere. Numerous strike-slip faults, along
which the crust is sliding sideways, cut the mountains – the Altyn
Tagh strike-slip fault forms the southern margin of the great Tarim
basin. North of the Tibetan plateau, the mountains decrease in
height – the Tien Shan mountains form some of the most northerly
of the Central Asian ranges.

long enough, it would be possible to detect changes in the shape of the landscape by comparing the angles in the triangles. This way one can measure directly the flow of the continents. The early surveys have now been compared with surveys made over 100 years later. The result is a picture of the movement of the landscape of New Zealand which varies smoothly from place to place like the flow of a fluid.

Many other mountainous parts of the continents – for example, in Greece and Turkey, California, the Andes and the European Alps – have now been measured in a similar way, but this time making use of the latest satellite surveying techniques. These studies confirm the general pattern of flow predicted by fluid models of the continents. In particular, they show that the flow is strongly influenced by the presence of high mountain ranges or depressions, which tend to drive the direction of flow, as we would expect in a fluid, from the high ground to the low ground. It is the surrounding push which keeps the mountains up, counteracting the tendency of mountains to flow downhill. If this push stops, the fluid theory predicts that, like a simple fluid, the mountains will flow away. There are parts of the world where this is happening today. In parts of western North America (called the Basin and Range Province because a whole series of normal fault lines have carved up the landscape into deep valleys and intervening ridges), and also in the Aegean of Greece and other parts of the Mediter- ranean, the mountains are flowing away, so that sometimes only the tops of high peaks are above sea level. In these regions, the Earth's crust is being stretched and thinned. Thus the fluid view of the continents suggests that mountains are ephemeral features.

Ultimately it is the movement of the plates which provides the push to keep the mountains up. So as the motion of the plates changes for whatever reason, the world's mountains will also change. Today's moun- tains will flow away, and new mountains may grow elsewhere in the world. It is the interplay between plate tectonics and the fluid-like behaviour of the continents which creates the distinctive mountain ranges around the world, each with their own particular history.

## RAISING TIBET

The fluid theory of mountains had explained many of the features of Tibet. But it faced a challenge when western geologists were first allowed to travel there in the early 1980s. Despite its extreme altitudes, Tibet is a land of lakes. Many of the lakes are elongated with a noticeable north–south orientation, filling deep valleys. The sides of the valleys are very steep and have all the tell-tale signs of fault lines. Geologists, examining these fault lines for the first time, discovered that these were faults that form where the crust is splitting apart – they marked great rifts in the Earth's crust. Some of these fault lines looked as though they had moved in geologically recent times, offsetting river beds and cutting deposits of rubble left behind by glaciers only a few tens of thousands of years ago. Seismologists began to report earthquakes in the region which also indicated that the crust was splitting apart. But there were more geological surprises. Geologists had spotted volcanoes in central and northern Tibet. The lava flows from some of these looked very fresh and recent. Again, the problems of getting into Tibet made the study of these volcanoes difficult, but the rare samples showed that the volcanic rock was coming from the mantle, beneath the crust.

Both the volcanoes and the rifts were not predicted by the fluid model. Philip England tried experi- menting with every conceivable situation, changing the properties of the fluid, changing the motion of India, but he could not produce a fluid-like Asia which was extending in an east–west direction, as indicated by the rifts in Tibet, or was hot enough in the mantle to start melting to produce volcanoes. This suggested that some other factor was at work which had been ignored in the fluid models.

*Right: Geysers of boiling water spurt out in the high Tibetan plateau. The hot water comes out near active faults where the crust is splitting apart.*

## MOUNTAIN ROOTS

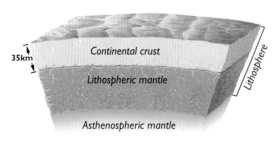

**(a)** Normal continental lithosphere

35km

Continental crust

Lithospheric mantle

Asthenospheric mantle

Lithosphere

**(b)** Mountains flow-up, 'iceberg' root formed

The continental lithosphere contains both crust and the top part of the mantle **(a)**. Mountains are pushed up when the continental lithosphere is compressed – in the process, both the crustal and mantle parts of the lithosphere are thickened, creating a deep root beneath the mountains **(b)**.

The mantle portion of the root is more dense than the underlying asthenosphere and eventually drops off, allowing the mountains to rise higher **(c)**. Hot asthenosphere flows in to fill the space left behind by the sinking blob, triggering volcanism; the high mountains begin to collapse and rifts form in the crust **(d)**.

**(c)** Bottom of lithosphere falls away, mountains bob up

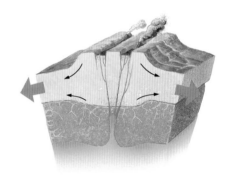

**(d)** Crust extended, volcanic activity

Here, we have to deal with a potentially confusing idea. The lithosphere, where the fluid flow that builds mountains is taking place, is not defined on the basis of composition but rather mechanical properties. It is composed of both crust and mantle. The lithosphere overlies a weaker portion of the Earth, the asthenosphere (see Chapter 4), which is also mantle – in other words, the base of the lithosphere is within the mantle. Because the temperatures in the Earth increase with depth, the lithospheric mantle is colder and more dense (the cold mantle contracts) than the hotter underlying asthenospheric mantle. This is called a density inversion and it is a gravitationally unstable situation – it is like having water on top of oil. Given a chance, the layering will reverse itself, and the oil will float above the water. If the material in a fluid-like continent is pushed up to form mountains, then the lithosphere, along with its crust, becomes thicker, and the overall instability of the density inversion between the two types of mantle increases. Eventually, the bottom of the lithosphere will literally drop off and sink back into the Earth's interior. This will cause the overlying continent to bob up locally, as the weight of the detached blob is removed (see above).

Philip England calculated that this effect could result in a geologically sudden uplift of the Tibetan plateau of between 1 and 3 kilometres. Once the overlying continent had bobbed up, it would tend to collapse and spread out because it would be much too high to be contained by the rest of fluid Asia. This collapse could explain the rifting seen in Tibet. The 'deblobbing' would be expected to have another consequence. It would expose a new bottom of the Asian lithosphere to the underlying hotter asthenospheric mantle. The latter could be hot enough to locally melt this region. The molten rock would rise to the surface and erupt in volcanoes. This could explain the presence of the young volcanic cones on the Tibetan plateau.

To some, Philip England's idea may be rather strange and complicated. We seem to have moved away from the overall simplicity of fluid continents that flow. But if one thinks about it, the 'deblobbing' process is just another manifestation of this flow. However, it points to a fundamental jerkiness in the flow. Instead of viewing mountains as features of the Earth which flow smoothly up or down, one might think of their growth in more catastrophic terms. Mountains may go up or down very rapidly. As we shall see, in doing so they may influence the workings of the planet in other ways.

## A CHANGE IN THE WEATHER

A major feature of the climate in Asia is the annual monsoon during the northern hemisphere summer. In northern India this usually starts in early July. In the weeks leading up to this, the air is almost unbearably hot and sticky. It comes as a relief when the torrential rain starts, bucketing down all along the Himalayas. Eventually, in September the rain eases up and the skies clear again.

Meteorologists agree that the intensity of the monsoon is to a large extent due to the presence of the high Himalayas and Tibetan plateau. During the summer months, the giant, high plateau is heated up by the Sun. Much of this solar energy is re-radiated, heating up the atmosphere over Tibet and making it

## TIBET, THE MONSOON AND LEAVES

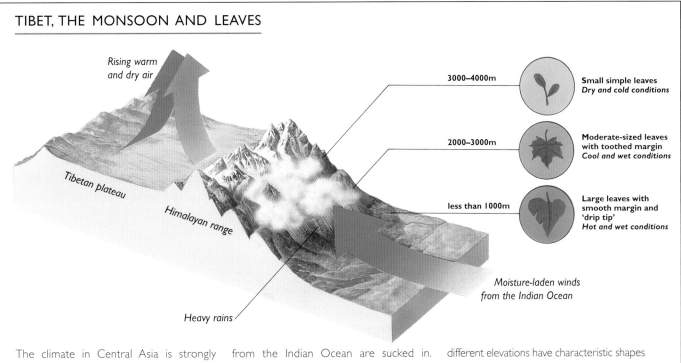

Rising warm and dry air

Tibetan plateau

Himalayan range

Heavy rains

Moisture-laden winds from the Indian Ocean

3000–4000m — Small simple leaves *Dry and cold conditions*

2000–3000m — Moderate-sized leaves with toothed margin *Cool and wet conditions*

less than 1000m — Large leaves with smooth margin and 'drip tip' *Hot and wet conditions*

The climate in Central Asia is strongly influenced by the presence of the Himalayan mountain range and the Tibetan plateau. During the summer monsoon, the high plateau warms up and the overlying air is heated. As the hot air rises, water-laden winds are sucked in. However, most of this moisture precipitates as rain over the Himalayas before reaching Tibet; for this reason Tibet is mainly a desert, while much of the Himalayas is cloaked in thick vegetation. Plant leaves at different elevations have characteristic shapes which are related to both the temperature and rainfall. Geologists can estimate the altitude of the mountains in the past by studying the shapes of ancient fossilized leaves now preserved in rocks.

*There are movements of the ground during an earthquake. By comparing radar images of California before and after the Landers earthquake in 1992, scientists have accurately determined these movements, shown as fringes of colour. The shape of the fringes can be thought of as contours linking areas that have moved by the same amount. Each colour cycle in the fringes represents about 2.8 centimetres of movement, increasing towards the fault-line which appears as a series of broken dark lines near the bottom right-hand corner. The image is about 80 kilometres across.*

significantly hotter than equivalent high parts of the atmosphere adjacent to Tibet. As the heated air rises, moist air from the Indian Ocean is sucked in to fill the void. But this moist air is blocked by the high peaks of the Himalayas, building up as clouds which eventually provide the heavy rain during the Indian monsoon (see p. 137). The Tibetan plateau also forms a barrier to a more general air flow, influencing the weather in a much wider region. Other high mountain ranges, throughout the world, also affect the weather. The high Andes on the western margin of South America form a barrier which deflects the airflow over most of South America, trapping heavy rain clouds in the Amazon basin. Even the relatively small ranges of the Southern Alps in New Zealand have an effect, focusing the rain on the west coast of South Island. These are not the only ways that mountain ranges can alter the climate, but this is a topic for another chapter (see Chapter 6).

This begs the question that if mountains can influence short-term weather patterns, what happens to the climate when new mountains such as the Tibetan plateau start to rise dramatically? Scientists believe that the intense monsoonal weather system we have today in India began about 8 million years ago. They can date this because, during the monsoon, the ocean near the Gulf of Aden is particularly rough, stirring up the sea bottom. This tends to increase the nutrients in this part of the ocean, resulting in an annual bloom of microscopic organisms. These eventually die and settle on the sea floor, year by year, layer by layer. Scientists studying the sea floor in this region have discovered that the annual bloom began about 8 million years ago. Could this also date a rapid rise of Tibet when the base of the lithosphere detached itself and fell into the underlying mantle? To test this requires knowing how the mountain range has grown through time.

Determining the height of a mountain in the past is notoriously difficult. Geologists have had to enlist the help of botanists, and a new field has developed called palaeo-altimetry. This is based on the way vegetation varies with altitude. If you walk up a mountain, you will notice that the vegetation gradually changes. The higher you go, the smaller the trees become until eventually you pass through the 'tree-line' and the mountain sides are just scrub, then bare rock. Botanists have detected detailed aspects of the leaves which also change progressively. For instance, at higher elevations the leaves become generally smaller and thicker with less complex shapes. One important reason for this is a change in temperature – the higher you go the colder it gets. Leaf shape is related also to the amount of rain the plant receives.

Botanists have used the variety of leaf shapes in a large collection of leaves to construct a 'leaf thermometer'. The thermometer can then be used to estimate altitude. This is particularly useful to geologists because it is based on features which can be found in the fossil record. The most likely place for leaves to be preserved is at the bottom of a lake. Leaves from the surrounding trees fall into the lake, leaving the imprint of their delicate structure on the muddy bottom. Sometimes the ancient remains of these lakes, now rock, are preserved. With care, the rock can be split parallel to the original layering, occasionally revealing almost perfectly preserved leaves.

Bob Spicer, a rare combination of geologist and botanist, has recently been to Tibet to collect fossil leaves. These are preserved in once extensive lake beds which existed in this region about 10 million years ago. The leaves suggest that they grew on trees in much warmer and wetter conditions than are found in Tibet today. Spicer thinks that the leaves are telling us that at this time Tibet was much lower, perhaps less than 4000 metres high compared to its average elevation of 5000 metres today. In other words, the leaves suggest that there was dramatic uplift of Tibet in the last 10 million years, just as Philip England predicted.

So it seems that an event, deep in the Earth's mantle, could have precipitated the raising of the highest region on Earth, triggering a change in the climate. The Himalayas became generally wetter, and the rivers cascading down these ranges transported huge quantities of rock debris to the Bay of Bengal in the Indian Ocean. This is the rock cycle, first discovered by James Hutton over 200 years ago (see

*Tibet today is a high, dry and very cold region. But a fossilized palm log, found in rocks a few tens of millions of years old, shows this was not always the case. This change in the local climate is mainly due to the uplift of the Tibetan plateau to its present average elevation over 5 kilometres above sea level.*

Chapter 1), operating with a vengeance. But the effects of the rise of Tibet may have been felt even further afield. Wetter conditions in Asia may have also resulted in drier conditions in East Africa, hastening the spread of the savannah lands and providing that window of opportunity exploited by our own distant ancestors when they left the forests and started walking on two legs. We will explore this in more detail in Chapter 7.

We have come a long way since those gold panning days of Harold Wellman on the west coast of South Island, New Zealand, over sixty years ago. We have changed our view of the continents, no longer thinking of mountains as ancient and fixed, but young and growing, fluid-like features of our planet. We have begun to see links between many aspects of our planet, from flow deep in the Earth's interior to the atmospheric circulation and climate. The fluid-like nature of mountains suggests that they are ephemeral things. Vast mountain ranges have been created and destroyed many times in Earth's history. Each time the atmospheric circulation has been affected. But, as we shall see in the next chapter, this is just one way in which the Earth's climate can change.

# CHAPTER 6

# THE ICE AGE

In the nineteenth century geologists discovered
evidence that large parts of the northern
hemisphere had once been covered by gigantic ice
sheets. Scientists have now learnt that the waxing
and waning of these ice sheets are just one aspect
of global climatic change, and that the planet has
been in the past both hotter and colder than it is
today. The complex interactions between variations
in the Earth's orbit around the Sun, the movements
of tectonic plates, the planet's atmosphere and
ocean currents, can result in large and rapid swings
in the Earth's climate.

*Today, large glaciers are found mainly in mountainous regions. This wide glacier is flowing
towards the sea in Alaska. The dark strip down the middle, called a medial
moraine, consists of rock debris which has been caught up in the ice.*

About 50 kilometres north of London, the Lea River flows through the rolling countryside of Hertfordshire. The river has cut into the underlying rock to form a wide valley with a flat bottom. Small depressions on the valley sides are the remains of quarries where farmers have dug down to the chalk bedrock for lime and flints. The sides of these old quarries are riddled with small holes – here rabbits have found that it is easy to burrow into a soft deposit on top of the chalk. A curious jumble of rock fragments, all the size of a fist, lies strewn around at the foot of the rabbit workings: ammonites, fossilized dinosaur bones, pieces of flint, and strange oyster shells called Devil's Toenails. There are also lumps of volcanic rock and granite which are usually only found far to the north in the Lake District or Scotland. If you look very hard you might also find the fossilized bones of elephants, hippopotami, sabre-tooth tigers, lions and woolly mammoths. So how did this jumble of fossils and rock types come to be among the rabbit burrows of Hertfordshire? The answer lies in violent changes in the Earth's climate – changes which have converted the now peaceful landscape near London from a tropical jungle to a frozen waste and back again many times in the past.

## A FROZEN WORLD

One hundred and fifty years ago the idea that the Earth's climate might have been radically different in the past seemed utterly fantastic to most scientists. And yet, even in historical times, there is evidence for climatic change. When the Viking Eric the Red sailed west of Iceland in the ninth century he discovered a fertile land he called Greenland. Subsequently, settlers began to farm this new country. But within a few hundred years the new colony was in trouble – the winters had become so severe that the farms and villages had to be abandoned. Europe witnessed a similar deterioration in the climate from 1450 until the late eighteenth century when the Baltic Sea and major rivers such as the Thames in England and Tagus in Spain began to freeze nearly every winter. Even in the last 200 years, villagers in the Swiss Alps have witnessed the Alpine glaciers advance and retreat during decades of unusual weather.

The strange deposit that lies near the surface of Hertfordshire can be found all over northern Europe

*When the ice sheet which covered northern Britain 18,000 years ago began to melt, large boulders which had been carried by the ice were left behind. This boulder, resting on limestone in Yorkshire, was plucked by the ice from bedrock much further west.*

*The base of an ice sheet is full of rock fragments. When the ice sheet advances, the rock fragments scratch grooves or striations in the underlying bedrock. These glacial striations in eastern Greenland are typical, and are a tell-tale sign of the presence of ice sheets in the past.*

and northern North America. When it was first recognized at the beginning of the nineteenth century, geologists had great difficulty working out what it was. It was often quite soft, forming a loose mixture of gravel, sand and clay draped over the landscape. Sometimes huge boulders of granite or some other exotic rock type, up to several metres across, were found embedded in the deposit or were left on the surface after the surrounding material had been eroded away. These boulders had polished surfaces or were marked with distinctive scratches and grooves. Drift, as this deposit came to be called, was clearly the result of a geologically recent event on a truly gigantic scale. The nature of this event was the subject of heated controversy in the early 1800s. The eminent British geologist William Buckland thought that this deposit was the remains of a global deluge – probably the biblical Flood.

European geologists like Jean de Charpentier and Jean Louis Agassiz had a different vision. Agassiz's

Over the last few tens of millions of years the Earth's climate has changed dramatically. The top painting is a reconstruction of the environment around Lucerne, Switzerland, 20 million years ago when conditions were tropical. But the same region, 18,000 years ago (shown below), was a frozen waste covered by large ice sheets, where mammoths roamed.

*The Lauterbrunnen valley in Switzerland was carved out during the last glacial period by a glacier which flowed from the high mountains in the distance. When the glacier retreated, a U-shaped valley with near vertical sides was left behind.*

inspiration came to him when he stood on a Swiss glacier in the 1830s. He noticed that the rubbly deposits which fill up the great valleys in the Alps, where there are glaciers, looked exactly like the drift. Loose boulders and the bare bedrock near these glaciers have the same distinctive scratches and grooves as those found in the deposit. Just by

watching the action of these glaciers it was obvious that the scratch marks were made as the glacier moved forward by the grinding action of rock fragments locked up in the ice. But to explain the wide distribution of the drift, Agassiz had to propose that at one time large parts of northern Europe and Scandinavia were covered by ice. When the ice

*Evidence for past Ice Ages can be found in the rocks. Here, scratch marks made by a glacier are preserved in Carboniferous rocks about 290 million years old, found near Kimberley in South Africa.*

melted, it left behind the drift. As geologists began to examine this deposit, they found that there was not just one layer, but several. The remains of Agassiz's icy world was interleaved with layers which contained the fossils of lions, hippopotami, elephants – creatures of the hot plains of Africa. So Agassiz was forced to the conclusion that the climate had changed rapidly, and not once but many times. Sometimes conditions were hot like Africa, sometimes they were cold like the Arctic regions.

Modern glaciologists have studied the drift in detail. It accumulates at the base of a glacier as the ice advances over the landscape. In the process, fragments of the underlying rock and older drift are plucked by the ice, ground up and incorporated into the ice sheet. The grinding action mixes up the rock types as well as generating a rock powder which becomes a clay matrix in the drift. Glaciologists have tracked the advances and retreats of the great ice sheets in the past by mapping the distribution of the drift and associated features. Even if the drift has been eroded away, or was never left behind at all, the rocky landscape can still have the distinctive mark of glaciers – polished and striated rock faces, hummocky hills sculpted by the advancing ice, or U-shaped valleys where the glacier has been confined between mountains and carved out its own valley.

Today, the great ice sheets in Antarctica and Greenland are extraordinary features of the planet – vast featureless landscapes of ice. The discovery of the drift has shown that ice sheets were even larger in the past. In the northern hemisphere, they formed a vast cap covering the north in Scandinavia, northern Europe, Siberia and northern North America, reaching as far south as New York. But periods in the planet's history when there are extensive ice sheets somewhere on Earth – periods which are called Ice Ages – have turned out to be surprisingly rare. We are in an Ice Age today but, as we shall see, this has only been so for the last 35 million years. The scattered remains of drift in the rock record show that the previous one was around 280 million years ago, lasting less than a few tens of millions of years. Even older drift deposits suggest major Ice Ages about 450 and 600 million years ago.

## A COOLING STORY

The record of past Ice Ages is scattered and fragmentary. But scientists now have a very good understanding of the sequence of events leading to the onset of the most recent Ice Age – the one we are in today. This is because the dramatic shift in the climate which was an inevitable part of the planet's slide into an Ice Age has left its mark on life. Living organisms are adapted to their environment. Any change in this will have an impact on them. For example, as we have already mentioned in Chapter 5, the shape of plant leaves is partly determined by the temperature of the environment in which the plants lived – geologists have used fossilized leaves to determine past average annual temperatures on land. Fossilized seeds or pollen of plants, preserved in sediments, reveal the type of vegetation – indicative of the climatic conditions – which grew in any particular region. However, good as these climatic indicators are, they are well preserved only in exceptional circumstances – for example, if there happens to have been a bog which was subsequently buried – usually they quickly decay away.

But there is another facet of living organisms which has provided an amazingly detailed and continuous record of climate change during the last hundred million years or so, the very period when the present Ice Age was born. The key to this is the oxygen locked up in fossils. Fossil shells are often made of calcium carbonate, a molecule which is partly composed of oxygen. The oxygen exists in several natural forms. By far the most abundant is a light isotope which weighs sixteen atomic mass units (oxygen-16). Next in abundance is a much less common heavier isotope which weighs eighteen atomic mass units (oxygen-18). During its lifetime, the creature that inhabited the shell slowly extracted oxygen from water to build its shell. Experiments have shown that the proportion of oxygen-16 and oxygen-18 in the shell is sensitive to the temperature. This is the temperature of the water in which the creature lived.

*Sediments on the sea floor contain a record of the climate over the past 100 million years. Sophisticated deep sea drilling technology is needed to extract long cylinders of rock from the sea floor; these samples are packed with information about past conditions in the ocean and atmosphere.*

## GLOBAL TEMPERATURES THROUGH TIME

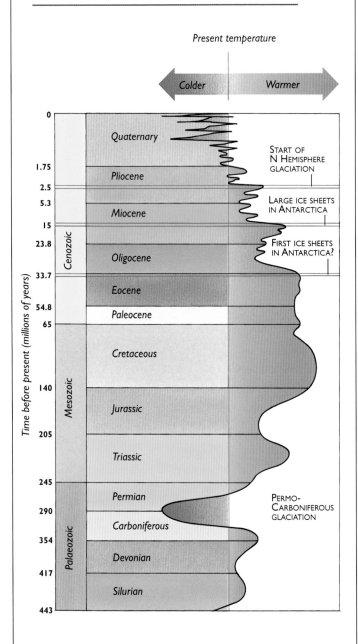

The mean global temperature has changed through time. This is revealed by the oxygen isotope record in deep sea cores, as well as the distribution of fossil plants and other geological evidence. Over the last 100 million years, the Earth has cooled from much warmer conditions than today, though there have been numerous small-scale swings in temperature. Large ice sheets started to grow in the last 35 million years. The previous time such large ice sheets existed was around 290 million years ago.

So scientists can estimate past ocean temperatures by determining the oxygen isotope ratios in the shells of microscopic ocean-dwelling creatures. The method is simple in theory, but extraordinarily laborious in practice. The shells have accumulated layer by layer, year by year, on the sea floor. Some are planktonic fossils, which had lived in the top 50 metres of the ocean. When these creatures died, their remains sank to join so-called benthonic fossils, which had lived on the sea floor. A slice through this pile of dead creatures is obtained by drilling into the ocean floor and extracting a cylinder of rock – thousands of these have now been sampled. Then begins the tedious business of identifying the layers and handpicking the different planktonic and benthonic species; this way both surface and deep water temperatures can be measured. The shells are analysed for their oxygen isotopes and the ratio of heavy and light oxygen is compared with the position along the core where the sample was picked – the periodic reversals in the Earth's magnetic field, which are recorded in the magnetism of the sediment layers, help to date these layers (see Chapter 2). When plotted on charts, the results produce squiggles, but these squiggles have turned out to be packed with information.

The record of ocean temperature extends back for several tens of millions of years. Surface ocean water temperatures have been fairly constant through time, though they have varied from ocean to ocean – the near surface water of more equatorial oceans is warmer than that for high latitude oceans, nearer the poles. However, deep ocean temperatures have undergone a remarkable change. About 100 million years ago, in the Cretaceous period when dinosaurs roamed the continents, the ocean bottom water was nearly 20°C warmer than today, and the oceans would have been like a tepid swimming pool. Today, the average ocean bottom temperature is nearly zero. The decline in ocean temperature reveals the big picture of climate change as the Earth moved into an Ice Age. Other climatic indicators show what was happening to temperatures on land. For example, plant remains suggest that in the Cretaceous, virtually everywhere, land temperatures were substantially warmer than today – there was no ice on land in either the Arctic or Antarctic, but instead huge tracts of forest.

But since this 'high' point in the climate, temperatures have declined overall, in a series of steps, to the conditions on Earth we know today. There was a marked cooling of water deep in the ocean around 35 million years ago. This coincides with the first signs of glacial deposits in deep sea sediments around Antarctica, suggesting that an ice sheet started to appear here on land. This marks the very early beginnings of the present Ice Age. About 15 million years ago, there was another pronounced drop in deep water temperatures, and the ice sheet in Antarctica was becoming well established and nearly its present size. However, as yet there is no evidence for an ice sheet in the northern hemisphere.

## THE PULSE OF CHANGE

At 15 million years ago something strange happens in the oxygen isotope record. Deep and shallow water fossils start to show rapid large and synchronized fluctuations in their oxygen isotopic composition. This is difficult to explain purely in terms of water temperature change, because surface and bottom waters do not usually change temperature in step. The explanation for this riddle has revealed a new way of reading the oxygen isotope record, which allows one to look outside the oceans at the ice sheets on land. To understand how, one needs to know about the distribution of water on Earth.

## OXYGEN ISOTOPES, ICE SHEETS AND SEA LEVEL

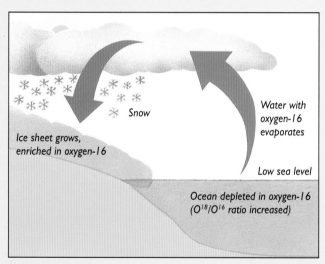

**(a)** Glacial period

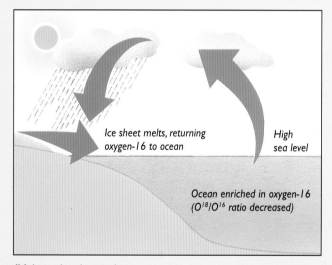

**(b)** Interglacial period

The waxing and waning of large ice sheets on land can affect the ratio of the light and heavy isotopes of oxygen in sea water. Water with light oxygen (oxygen-16) evaporates more readily from the oceans. Some of this water vapour will eventually freeze as snow and accumulate on ice sheets during the winter months. During a glacial period, the snow does not melt much during the summer and the ice sheets grow, taking water out of the oceans and lowering the sea level **(a)**. This way, the water in the ice is enriched in light oxygen, but the oceans are depleted (i.e. the ratio of heavy to light oxygen in oceans increases). During an interglacial period the ice sheets melt. The melt water raises the sea level and enriches the oceans again with oxygen-16 **(b)**.

Water is found in many parts of the planet: some is locked up in rocks at great depth, but the rest is either very near or on the surface as liquid water, ice and snow, or as moisture in the atmosphere. Today, nearly 97 per cent of the surface water is in the oceans. Most of the rest is in glaciers as ice (nearly 3 per cent), and a much smaller amount is in rivers, lakes, soils, water aquifers, the atmosphere and living organisms. Thus the oceans and glaciers contain almost all the surface water on Earth. Today, the large ice sheets in Greenland and Antarctica grow through the accumulation of snow at high latitudes. The snow forms during the winter months when moisture in the atmosphere freezes. Most of this moisture comes from the evaporation of water in the oceans. So the growth of ice sheets and glaciers can be seen as a transfer of water from the oceans to the land. If more water is lost to the atmosphere by evaporation than is returned by

rivers and melting ice, then the volume of the oceans will decrease, sea level will fall, and the volume of ice on land will increase. If, as appears to be happening today, more water is flowing into the oceans than is being lost through evaporation, then the volume of the oceans will increase, sea level will rise, and the ice sheets and glaciers will shrink.

Water, of course, is hydrogen and oxygen ($H_2O$), and the waxing and waning of ice sheets affects the ratio of oxygen isotopes in the oceans. This is because the evaporation 'pump' which transfers water from the oceans to ice sheets filters out some of the heavy oxygen – water molecules made with light oxygen evaporate much more readily than water with heavy oxygen. So, during periods when the ice sheets on land grow, the oceans are selectively depleted of light oxygen. When the ice sheets shrink, melt water is returned to the oceans and the proportion of light

## OXYGEN ISOTOPE RECORD IN DEEP SEA CORES

Fossil shells, buried beneath the sea floor, contain a record of past conditions in the oceans. By measuring the relative amounts of the light and heavy isotopes of oxygen locked up in the shells, it is possible to calculate both the temperature of the ocean water in which the fossil creature lived and the volume of ice sheets on land. The record in deep sea cores from the Pacific suggests that about 2.5 million years ago the climate began to cool markedly and ice sheets became extensive in the northern hemisphere (a). The record for the last 250,000 years shows that there have been numerous advances and retreats of the ice sheets which correlate with changes in sea level (b–c).

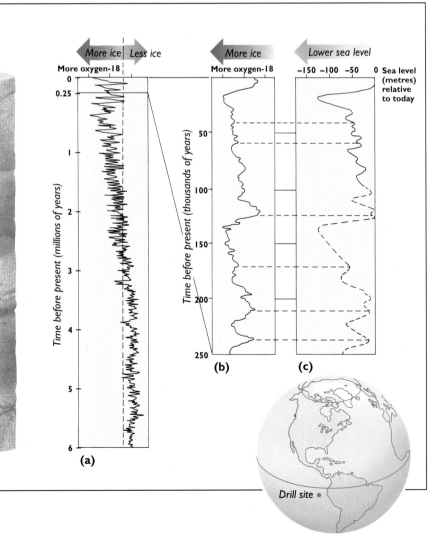

oxygen in them increases again. Therefore, changes in ice sheet volume through time will result in fluctuations in the proportion of light and heavy oxygen in sea water; this will be reflected in the oxygen isotope ratios in fossil shells of organisms in both deep and shallow water.

When there is not much ice on land, this does not have much effect. But it seems that about 15 million years ago there were ice sheets sufficiently large for this effect to swamp that of water temperature change. So the oxygen isotope ratios for this period should be read in terms of changes in ice volume rather than temperature. The record is particularly clear over the last six million years – there have been times when much more of the world's water was locked up in ice on land, and sea level was low. These have alternated with times when the ice volume was small, and sea level was high (see opposite).

About 2.5 million years ago, the oxygen isotope record suggests that the ice volume on land started to increase substantially (see opposite). Deep sea sediments in the North Atlantic Ocean contain for the first time the tell-tale signs of ice sheets in the northern hemisphere – isolated 'drop' stones which have fallen from icebergs. For the icebergs to carry rock, they must once have been part of a continental ice sheet, before breaking off and floating out to sea. This marks the beginning of the northern hemisphere glaciation. The ice sheets which started growing at this time have left behind the drift – the deposit which first attracted geologists to the notion of Ice Ages. In the last million years there have been several phases of substantial growth of the northern hemisphere ice sheets, separated by times when they have retreated; these are the advances and retreats which are revealed by the drift.

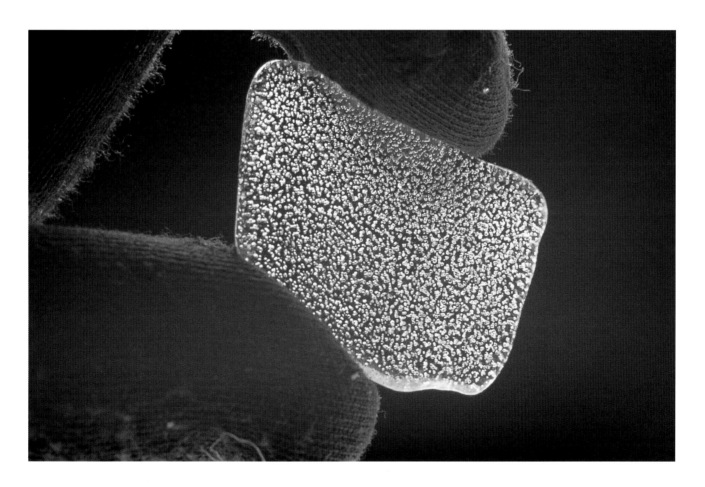

*This ice formed nearly 200 years ago in Antarctica. Natural ice is full of small gas bubbles. The bubbles are samples of the atmosphere, trapped in the ice – scientists can study the composition of the atmosphere in the past by analysing the gases.*

## PORTRAIT OF A GLACIAL PERIOD

The sediment cores provide a remarkably continuous and precise record of the periodic growth and retreat of the world's ice sheets over the past 2.5 million years. Yet, astonishingly, scientists have discovered an even more detailed record of climatic change, one which allows them to reconstruct the past few hundred thousand years almost year by year.

Every summer scientists from all over the world converge on the vast Greenland ice cap. Here, 3000 metres above sea level and thousands of kilometres from the nearest human habitation, the scientists are drilling into the ice.

The proportion of light and heavy oxygen isotopes in snow reflects the average temperature of that level in the atmosphere where moisture freezes as snow. Thus oxygen isotope ratios can also be used as a way of measuring past atmospheric temperature. Fresh layers of snow accumulate year by year and are gradually compressed into ice. By drilling into the ice, scientists can uncover the layers once more – dating the ice is a matter of counting back the annual layers. Occasionally, dated fall-out from volcanic eruptions, trapped in the ice, acts as markers.

The ice cores show that the average annual temperature in Greenland has oscillated violently in the past. If these temperature variations are plotted on a chart they appear as squiggles (see p. 154). During the last 10,000 years – geologists call this period the Holocene – the temperature has varied frequently, but not by very much; the squiggles are quite small. But going back beyond the Holocene, it is a very different story. There have been rapid swings in temperature up to 10°C every few thousand years. However, the lowest temperatures were substantially colder than anything during the Holocene. It is only if one goes back in time more than 100,000 years that temperatures are

*The edge of the vast ice sheet in east Antarctica. The ice on the horizon rises to over 3000 metres above sea level. During the last glacial maximum, much of northern North America and Europe was covered by ice sheets like this.*

the same as, or even warmer than, those in the Holocene.

Geologists have put together the evidence from both deep sea and ice cores to paint a picture of the growth and decay of the giant ice sheets; this appears in the deep sea oxygen isotope record as a saw-tooth pattern, which they call a glacial cycle. To appreciate this it is helpful to take the situation today as our point of reference. Today, we are in what is called an interglacial period, when ice sheets exist but are small in the northern hemisphere, sea level is high, and polar temperatures are warm. The previous interglacial period when, incidentally, temperatures were sometimes much warmer than today – hence evidence for hippos and elephants in Hertfordshire – ended about 115,000 years ago, and a glacial period began. Subsequently, as high latitude temperatures decreased, ice sheets began to grow larger in the northern hemisphere, and also in the southern part of South America (the ice sheet in Antarctica was already extensive), and sea level fell. This overall trend was interrupted by a number of brief phases of

## OXYGEN ISOTOPES IN GREENLAND

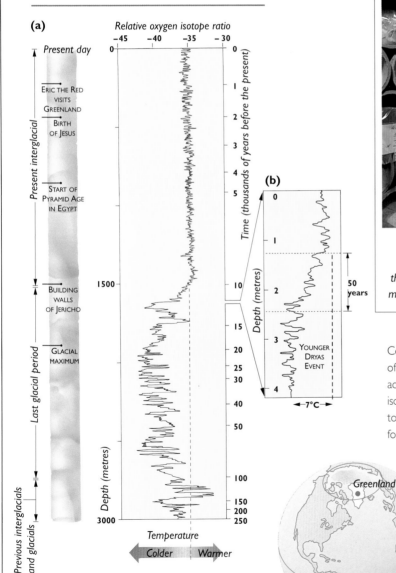

*The ice in Antarctica and Greenland accumulates year by year. Scientists can study the Earth's climate in the last few hundred thousand years by analysing the annual layers of ice. To do this, they must drill into the ice sheets and extract long cylinders (cores) of ice.*

Cores of ice, extracted from the Greenland ice sheet, contain a record of the climate over the last 250,000 years **(a)**. Over time, the ice has accumulated. By measuring the relative amounts of the light and heavy isotopes of oxygen in the ice at particular levels in the core, it is possible to calculate the atmospheric temperature when the ice at this level formed. During the present interglacial period, in the last 10,000 years, the temperature has been fairly constant, but in the period before that, in the last glacial period, the temperatures have been generally colder, swinging rapidly back and forth. For example, about 11,500 years ago, the temperature increased extremely rapidly at the end of the so-called Younger Dryas Event **(b)**.

# ICE SHEETS IN THE NORTHERN HEMISPHERE

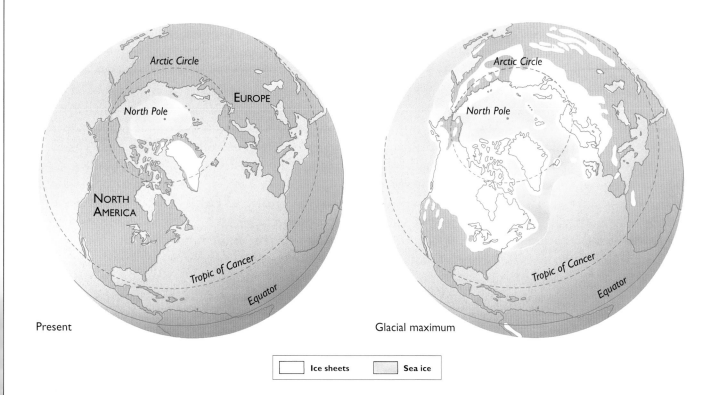

Present

Glacial maximum

Ice sheets    Sea ice

Today there are only two large ice sheets on land – in Greenland and in Antarctica. In the last million years during glacial maxima – for example, 18,000 years ago – there were extensive ice sheets, several kilometres thick, covering much of northern North America, northern Europe, Scandinavia, Finland and northern Siberia. Year-round sea ice, which is limited today to the polar regions, extended as far south as British waters and the eastern seaboard of North America.

warming, called interstadials, when the ice sheets retreated and sea level rose again slightly. By 18,000 years ago, the glacial maximum had been reached, when ice sheets were at their greatest extent and sea level was lowest (see above). Soon afterwards, as the climate warmed again, the ice sheets began to shrink rapidly and sea level started to rise to its present level. Judging by the past trend, the start of cooling which will take us into the next glacial period is only a few tens of thousands of years in the future.

An international project called CLIMAP has come up with a detailed picture of the Earth in the month of July during the last glacial maximum, 18,000 years ago, using a wealth of geological information. The month of July was chosen because the long-term survival of northern hemisphere ice sheets is very sensitive to temperatures during the summer months. At today's summer temperatures, the winter snow accumulation largely melts and the ice volume either remains the same or shrinks. The oxygen isotope record suggests that, on average, surface temperatures were lower everywhere 18,000 years ago, so the mean global temperature would have been several degrees less than today. Temperatures were particularly low over the vast ice sheets that covered much of the northern part of the northern hemisphere. The winter snow lasted during the cool summer months. In contrast, surface temperatures near the equator in Africa were only very slightly lower (about 2 to 3°C lower). Studies of fossilized pollen from plants show that, like today, there were tropical forests near the equator, but conditions were generally drier. From an equatorial point of view, climate change into and out of a glacial maximum seems to be a contraction and expansion of the tropical zone, rather than its complete obliteration.

## THE RISING SEA

As we have seen, the growth of ice sheets takes water from the oceans. The inevitable consequence of the glacial cycle is periodic changes in sea level around the globe. The islands of New Zealand and New Guinea contain an unusually good record of these sea level changes. The coastal landscape has a characteristic shape made up of a series of terraces at progressively higher levels. In places, these terraces are so regular that the landscape looks almost like a giant flight of stairs rising up from the sea. Each terrace is a step on this staircase, and at the back of each step, where the cliffs rise up to the next step, the remains of an ancient beach can be found.

Charles Cotton, a New Zealander living near the turn of the century, worked out how the terraces form.

Imagine a situation when sea level is high. Powerful waves along the exposed New Zealand coastline eat into the land, cutting cliffs and creating a rocky platform which is exposed at low tide. Over time this rocky platform gets wider and wider, as coastal erosion progressively undermines the cliffs. Then, if some time later, the sea drops dramatically several tens or hundreds of metres, the platform will be left stranded high and dry above the new lower sea level. If the sea rises again, it will begin to flood the coastline. However, New Zealand is a land of earthquakes, and is being pushed up, squeezed between the Pacific and Australian plates. Thus the coastline has been steadily rising out of the sea and the old rocky platform will be out of reach of the rising flood, preserved as a raised terrace near the coast, and the waves will begin to cut a new rocky platform at a lower level than the old terrace. This sequence of events, repeated many times, will create the staircase of raised terraces (see opposite).

*The coastline of North Island, New Zealand, is slowly rising out of the sea, pushed up by local Earth movements. In the process, a series of terraces, which were formed during past changes in global sea level, have been preserved. A prominent high terrace is visible in the foreground, and a lower terrace forms a coastal plain.*

## THE CREATION OF COASTAL TERRACES

Distinctive terraces often form along coastal regions which are being raised by Earth movements – these terraces are a record of global sea level change. The sequence of events which creates the terraces is shown here.

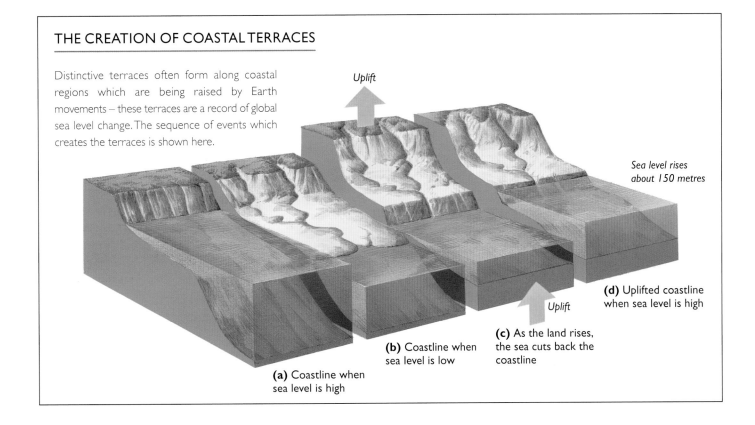

*Uplift*

*Sea level rises about 150 metres*

*Uplift*

**(a)** Coastline when sea level is high

**(b)** Coastline when sea level is low

**(c)** As the land rises, the sea cuts back the coastline

**(d)** Uplifted coastline when sea level is high

Coastal platforms are also created when sea level is low, but these tend to be drowned when the sea rises and so are not preserved on land. However, the record of low stands is sometimes preserved offshore. In Barbados, corals which grow only within a few metres of the sea surface are now found tens of metres below sea level. Dating these corals is the way to document a detailed history of the most recent rise in sea level – those nearly 18,000 years old are over 100 metres below sea level, and younger ones are at progressively shallower depths. The rise, overall, was about 120 metres, starting about 18,000 years ago, and in some places the shore line advanced over the coastal plains as much as a kilometre a year.

Throughout the world, detailed surveys of the shallow continental shelves have revealed the traces of many of the old shorelines. In fact, the continental shelves themselves are the work of rivers when sea level was much lower. Then, the rivers had to flow farther to reach the sea. The detritus they deposited has built up to form a wide coastal plain which was flooded when sea level rose. The depths of the continental shelves are remarkably uniform throughout much of the world, being between 100 and 200 metres below present sea level. This depth is a measure of the height difference between low and high stands of sea level; the continental shelves formed wide coastal plains at low stands and shallow shelf regions at high stands, like today (see p. 159).

Modern humans were already hunting and fishing 18,000 years ago, as do many tribal peoples today. The world-wide flooding of coastal regions during the most recent rise in sea level may be the event behind the flood legends which exist in so many different cultures. Such catastrophes cause large-scale displacements of peoples, and are rarely forgotten, though the details may be distorted. For instance, the Book of Genesis and the Assyrian Epic of Gilgamesh both contain accounts of a catastrophic global flood. The Australian Aborigines also have a flood story which may be a distant echo of the flooding of the continental shelf region which today separates Australia from New Guinea. The Aborigines record in songs and dances the sea encroaching on their hunting grounds.

## UNDERSTANDING THE CLIMATE

So far we have documented the evidence for dramatic climate changes which accompanied the Earth's passage into the present Ice Age. The evidence shows that these changes were sometimes rapid and repetitive. So why are there Ice Ages and what determines the variation within one?

We can think of these climatic fluctuations in a number of ways. In broad terms the surface of the Earth is cooler during an Ice Age than out of one. Scientists have made thousands of measurements at sea and on land to determine the average surface temperature of the Earth today – this is roughly 15°C.

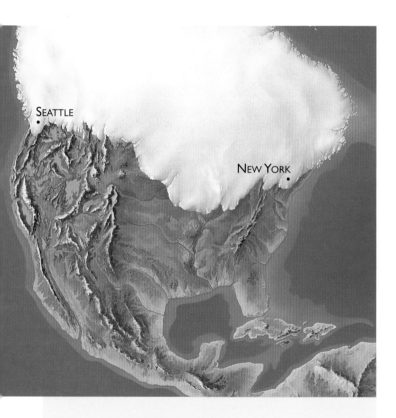

## GLACIAL NORTH AMERICA

There have been times during the last million years when ice sheets covered much of northern North America, extending as far south as New York on the east coast and Seattle on the west coast. South of the great ice sheet there were mountain glaciers and large lakes.

During the mid Cretaceous, when the Earth was not in an Ice Age, this temperature was several degrees higher. But, in detail, there is a temperature gradient between the poles and equator; the poles are always colder than the equator, and ice sheets form mainly at the poles (see p. 160). The history of the Earth's climate has shown that both the average surface temperature, and indeed, as the CLIMAP project has revealed, the temperature gradient between the poles and equator, have varied over time.

Climate change is simply the sum of long-term changes in the atmosphere. But trying to make sense of all aspects of climate change is virtually impossible without some coherent model of the atmosphere with which to assess their significance. The model exists inside a computer. It is mainly concerned with the region within 30 kilometres of the Earth's surface, called the troposphere, where virtually all of the mass of the atmosphere resides. This is in continual motion, heated almost entirely by the Sun's rays – a source of energy 1000 times more powerful than the heat which flows from the Earth's deep interior. The radiation is much more oblique near the poles, and nearly head-on at the equator; this results in a pole to equator temperature gradient. There is also a profound interaction between the atmosphere and both the oceans and land surface. Physicists call the atmosphere a non-linear system, because a small change in one part can have a knock-on effect disproportionate to the original change, sometimes leading to catastrophic results. They consider the temperatures and pressures of the gases, which behave according to Newton's laws of motions. Today, the best computer models cannot deal with weather or climate patterns which are at a smaller scale than 300 kilometres; to place this in context, a typical thunderstorm is only 50 kilometres across. Computers a thousand times faster than the fastest computers available today would be needed to model such smaller-scale features of the atmosphere. Despite these limitations, the computer model has proved to be an extremely powerful weapon in the armoury of climatologists as they struggle to study climate change. It has highlighted the factors which they need to consider.

## MILANKOVITCH CYCLES

The behaviour of the atmosphere is to a large extent controlled by the amount of solar radiation that falls on Earth. But the intensity of this radiation varies with the changing position of the planet as it orbits the Sun. In the 1920s, a Yugoslavian mathematician called Milutin Milankovitch showed that the orbit varies in three distinct ways (see p. 162).

The path of the Earth about the Sun is an ellipse. The shape of this orbit changes over time, varying from a more elliptical to a more circular path and back again every 100,000 years or so. On a shorter timescale, the orientation (obliquity) of the Earth's spin axis, relative to the plane of the Earth's orbit, also oscillates between two extreme positions. This happens more rapidly than changes in the ellipticity of the orbit, with a 41,000-year cycle. Finally, in detail, the spin axis wobbles like a spinning top as it precesses. These wobbles, combined with a slow drift of the elliptical orbit of the Earth, cause a steady shift in the date when the Earth is closest to the Sun. The precession of the equinoxes, as this is called, takes 19,000 to 23,000 years before the cycle is completed.

Milankovitch realized that the various wobbles in the Earth's orbit will affect the distribution and intensity of the incident solar radiation which reaches the Earth's surface. He thought that the global climate was particularly sensitive to the incident radiation

## REGIONS FLOODED DURING INTERGLACIALS

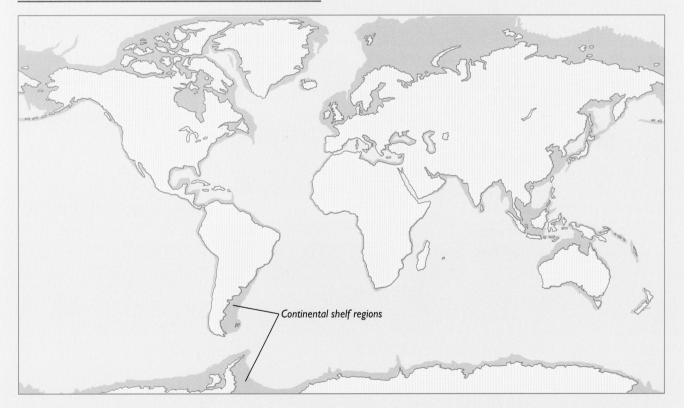

Continental shelf regions

During the last million years the Earth's climate has oscillated between glacial conditions, when large ice sheets existed on land, and warmer interglacial periods. At the end of each glacial period – for example, approximately 18,000 years ago – ice sheets began to melt and the sea level rose, flooding wide coastal plains. Today these drowned plains form the continental shelves (shaded in green on this map) which are between 100 and 200 metres deep – their depth reflects the rise in sea level.

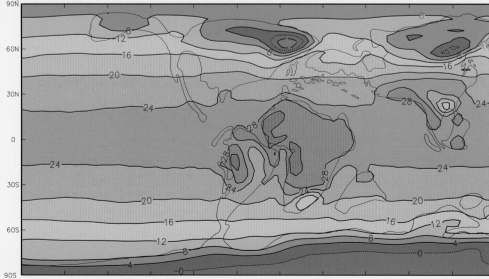

## TEMPERATURE OF THE EARTH'S SURFACE

The average annual surface temperature varies over the surface of the Earth – the equator is hotter than the poles. Climatologists have reconstructed this temperature distribution in the past (shown here contoured in °C from warm – red to cold – purple), using a variety of temperature indicators preserved in the geological record, combined with computer models of the Earth's atmosphere. This work shows that about 100 million years ago, in the Cretaceous period, the Earth was warmer overall than today, and forests existed at the poles. During the last glacial maximum, 18,000 years ago, the Earth was overall a few degrees colder than today – temperatures at the equator were only slightly lower, but the polar regions were much colder.

Today

Last glacial maximum, 18,000 years ago

Mid-Cretaceous, 100 million years ago

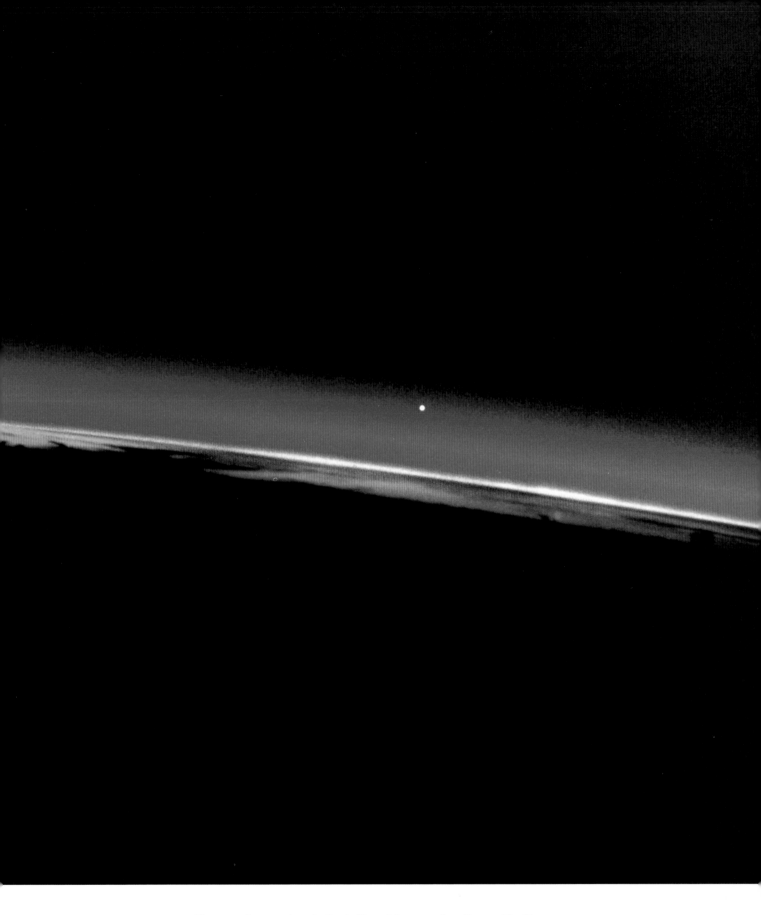

The atmosphere envelops the Earth. Most of the atmosphere forms a thin skin
only about 30 kilometres thick, catching the light in this space shuttle image.

# MILANKOVITCH CYCLES

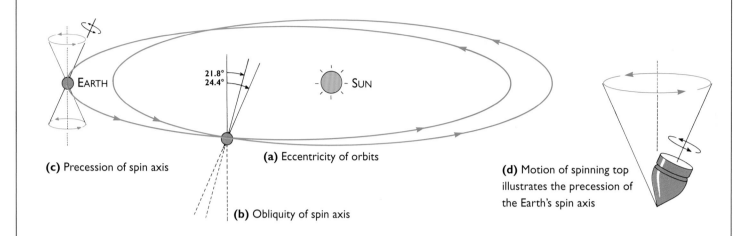

**(c)** Precession of spin axis

**(a)** Eccentricity of orbits

**(b)** Obliquity of spin axis

**(d)** Motion of spinning top illustrates the precession of the Earth's spin axis

The orbit of the Earth around the Sun varies through time in three ways. The Earth follows an elliptical path; however, the shape (eccentricity) of this changes, from more elongate to more circular **(a)**. In addition, the orientation (obliquity) of the Earth's spin axis fluctuates relative to the plane of the Earth's orbit **(b)**. Finally, in detail, the spin axis precesses like a spinning top **(c–d)**. All these factors affect the amount of solar radiation reaching the Earth's surface, and therefore the climate, in characteristic cycles called Milankovitch cycles.

which arrives at mid to high latitudes. Milankovitch laboriously produced a series of curves showing how the incident solar radiation here had varied over the last few hundred thousand years. When Milankovitch first published his results, before the Second World War, past climate change was too poorly known to provide an adequate test of his idea. Climatologists disputed whether these astronomical effects were large enough to provoke a significant change in the climate.

In the 1970s, scientists around the world were able to directly compare the climatic oscillations revealed by the oxygen isotope record in deep sea cores with the variations in the amount of solar radiation calculated by Milankovitch. It was clear that the Milankovitch cycles, as Milankovitch's curves came to be known, explained at least 60 per cent of the observed climatic fluctuation. This has been taken by most scientists as vindication of Milankovitch's astronomical theory, providing an explanation for the driving force behind much of the climatic variation over tens and hundreds of thousands of years on Earth.

The deep sea cores contain a record of growing and decaying ice sheets. So Milankovitch cycles, resulting in variations in the incident solar radiation,

must be influencing the size of the ice sheets. Climatologists are now, with the benefit of their computer models, beginning to see how this might be so. The variation in incident solar radiation, calculated by Milankovitch, must directly affect the temperature contrasts between seasons. Thus, if seasonal variation is reduced at the mid to high latitudes, such that the summers are slightly cooler, then there will be less summer thawing – ice sheets will grow. Greater seasonal variation, in particular hotter summers, will result in more summer melting and a general shrinking of the ice sheets.

One puzzling feature of the Milankovitch cycles is that the weakest astronomical forcing of the climate has resulted in the strongest climatic signal over the past million years – the cycle between glacial and interglacial periods every 100,000 years or so. Small changes in the ellipticity of the orbit on the 100,000-year timescale will affect both the amount of solar radiation which reaches the Earth during particular seasons and the total over the year. The intensity of the solar radiation is inversely proportional to the square of the distance from the Sun, in the same way that a torch beam appears dimmer when shone upon distant objects. In fact, the changes in the ellipticity of the Earth's orbit round the Sun are very small, causing less than a 0.3 per cent fluctuation in the intensity. To explain how this can push the planet into or out of a glacial, one has to consider the non-linear nature of the atmosphere, so that small effects are amplified and result in large changes in the climate.

## LIVING IN A GREENHOUSE

Milankovitch cycles are an example of a force outside the Earth which can disturb the climate. They explain climate change within an Ice Age, but not why Ice Ages exist in the first place. There are equally powerful agents of climate change *within* the atmosphere. Particular gases in the atmosphere have a profound effect on the temperature of the Earth's surface; the most important today are water vapour, carbon dioxide, methane, nitrous oxides, chlorofluorocarbons (CFCs) and ozone. The computer climate models suggest that without these gases, the average temperature of the Earth's surface today would be $-6°C$ instead of $15°C$. If their level rises, then, in general, this temperature will also rise, and vice versa. This is known as the greenhouse effect, because the so-called greenhouse gases act in very much the same way that a greenhouse maintains a warm environment for cultivating plants, forming a sort of giant thermal blanket. It is the observed massive increase in the concentration of carbon dioxide in the last 150 years which is now thought to be the prime cause of global warming in our lifetime. In fact, the computer models of the atmosphere suggest that the predicted doubling of the present concentration of greenhouse gases in the next fifty years will result in an increase in the average surface temperature of about $2°C$.

Russian scientists have extracted a remarkable record of levels of the greenhouse gases such as carbon dioxide and methane in the atmosphere over Antarctica during the last 150,000 years (see p. 164). They have done this by drilling into the ice at their Vostok base and analysing the gases trapped in minute bubbles in the ice; these bubbles are thought to be samples of the air at the time when the ice at any particular level in the ice core formed (see p. 164). By analysing the oxygen isotopes in the ice, they have also determined atmospheric temperature through time at this site. These records show a striking correlation between the level of carbon dioxide and methane and temperature. This result prompted the notion that the level of greenhouse gases regulates the planet's climate over even longer timescales, extending for tens and hundreds of millions of years; the Earth alternates between so-called greenhouse and icehouse states, when the atmosphere is either warm or plunged into colder Ice Age conditions. However, the rarity of Ice Ages over the history of the Earth suggests that if this is the case, then the greenhouse state is the normal one which is only occasionally perturbed into icehouse conditions.

Robert Berner at Yale University has attempted to test this theory by calculating the level of greenhouse gases in the atmosphere through time. Unfortunately, there is no direct measure of this in the geological record prior to about 150,000 years ago, and so Berner has tried to track down all the parts of the Earth where the chemical components of these gases may reside, concentrating on the fate of the important greenhouse gas carbon dioxide. It turns out that there are a limited number of ways that carbon dioxide can be exchanged between the solid parts of the Earth and the atmosphere. Ultimately, it seems to be both biological activity and convection in the Earth's mantle, expressed at the surface as the motion of the tectonic plates and volcanic eruptions, which control the level of carbon dioxide in the atmosphere.

Living organisms extract carbon dioxide from the atmosphere as part of the chemistry of life. If organisms are buried rapidly when they die, there will be a net loss of carbon dioxide from the atmosphere. Eventually, this carbon dioxide may be returned, especially if the organic remains are heated up or are exhumed during earth movements. Erupting volcanoes spew out large quantities of carbon dioxide, ultimately derived from the Earth's mantle. A reversible process known as the Urey reaction (named after the Nobel laureate who first described it) has an important influence on the levels of carbon dioxide

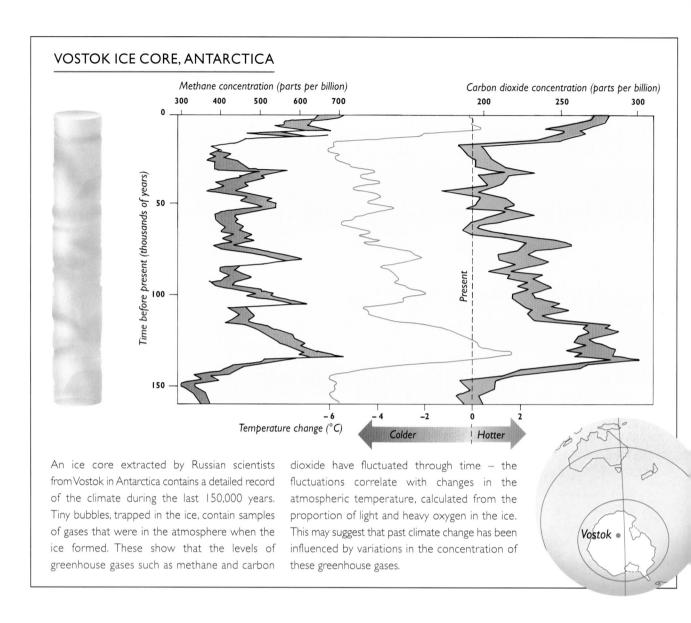

## VOSTOK ICE CORE, ANTARCTICA

Methane concentration (parts per billion)

Carbon dioxide concentration (parts per billion)

Time before present (thousands of years)

Present

Temperature change (°C)

Colder    Hotter

Vostok •

An ice core extracted by Russian scientists from Vostok in Antarctica contains a detailed record of the climate during the last 150,000 years. Tiny bubbles, trapped in the ice, contain samples of gases that were in the atmosphere when the ice formed. These show that the levels of greenhouse gases such as methane and carbon dioxide have fluctuated through time – the fluctuations correlate with changes in the atmospheric temperature, calculated from the proportion of light and heavy oxygen in the ice. This may suggest that past climate change has been influenced by variations in the concentration of these greenhouse gases.

in the atmosphere. Carbon dioxide, dissolved in rain water, can react with silicate rocks on the Earth's surface. The end-products of these reactions are magnesium or calcium-rich limestones which, in effect, contain carbon dioxide extracted from the atmosphere. If the limestones are subsequently buried deep within the Earth's crust and heated up, the process is reversed and carbon dioxide is released back into the atmosphere

Robert Berner's calculations show a persuasive link between long-term climate change and atmospheric carbon dioxide levels. It has been suggested that the Ice Age near the end of the Precambrian, between 700 and 600 million years ago, was the result of a global cooling caused by a massive drawdown of carbon dioxide when single-cell organisms started to extensively colonize shallow seas. Subsequently,

carbon dioxide levels peaked at over ten times today's level, before returning to levels comparable to today around 280 million years ago when extensive ice sheets covered parts of the supercontinent of Gondwanaland, comprising what are today the southern hemisphere continental masses.

The level of carbon dioxide about 100 million years ago was perhaps five times greater than that today. This coincides with a particularly warm period in the Earth's history when large cold-blooded dinosaurs dominated the land and there were forests at the poles. Since then, there has been the steady cooling of the climate already described earlier in this chapter, which seems to coincide with a general decline in atmospheric carbon dioxide levels – a reduction in the atmospheric levels of this important greenhouse gas would be expected to cause global cooling. At the same time, plate motions have been pushing up the vast ranges of the Himalayas and Tibet in Central Asia, and the Andes in South America (see Chapter 5). The American climatologists Maureen Raymo and William Ruddiman think this is not merely a coincidence – they have suggested that as a consequence of the Urey reaction, taking place during the large-scale erosion and weathering in these mountains, carbon dioxide has been sucked out of the atmosphere. The process continues as more rock is pushed up during mountain building, acting as a sort of pump transferring atmospheric carbon dioxide from the atmosphere to the weathering products of rock. These are washed into the oceans where they remain for tens of millions of years. Today, as part of our widespread industrial activity, we are returning large quantities of carbon dioxide to the atmosphere which was extracted by living organisms hundreds of millions of years ago. We may therefore be causing the present global warming.

## A TWIST IN THE TAIL

Many climatologists are convinced that something is missing from the greenhouse theory of climate change. One problem is that the atmosphere operates as a coupled system, linked to so many aspects of the surface of the Earth, that it is often difficult to sort out cause and effect. It is quite possible that in some cases it is global cooling that first triggers a decrease in the level of greenhouse gases, such as carbon dioxide, in the atmosphere, rather than the other way round. This can come about because the cooling of the atmosphere is more rapid than the cooling of the oceans, so atmospheric cooling results in an increase in the temperature gradient in the near surface parts of the ocean. This sets up more vigorous currents which stir up more of the top part of the ocean. The

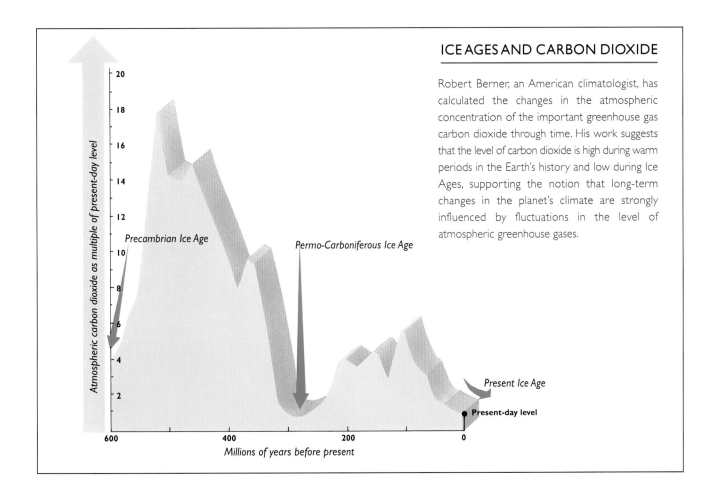

### ICE AGES AND CARBON DIOXIDE

Robert Berner, an American climatologist, has calculated the changes in the atmospheric concentration of the important greenhouse gas carbon dioxide through time. His work suggests that the level of carbon dioxide is high during warm periods in the Earth's history and low during Ice Ages, supporting the notion that long-term changes in the planet's climate are strongly influenced by fluctuations in the level of atmospheric greenhouse gases.

stirred-up water is better oxygenated with a greater distribution of nutrients, promoting blooms of small planktonic organisms. These take up more carbon dioxide as they grow, and so a biological pump is set in motion which will draw down carbon dioxide from the atmosphere, promoting further global cooling.

There are a number of other little wrinkles in the greenhouse theory. For instance, the long-term history of cooling during the last 40 million years, if ultimately caused by the carbon dioxide draw-down effect of the Urey reactions in the Himalayas, actually requires some additional process which returns carbon dioxide to the atmosphere; the Urey reactions are so effective that on their own they would have stripped the atmosphere of carbon dioxide long ago! Another problem arises when the effects of greenhouse gases are examined in computer models. These models tend to predict a general warming of the climate at all latitudes when the concentration of greenhouse gases increases, while the available climatic indicators for past warm periods, such as fossil plants and the oxygen isotope record, suggest a more complex picture. They consistently show warming to be greater towards the poles, but marginal near the equator, in the tropics. This suggests that there is another key factor in climate change. Perhaps it is the configuration of the ocean current system, which is so effective at redistributing heat over the surface of the globe?

Ocean currents can influence the growth of ice sheets by modulating the atmospheric temperatures in critical latitudinal zones. There is a tendency for equatorial warm water near the surface of the oceans to flow in great gyres towards the poles, driven by the winds. The Gulf Stream brings warm surface waters from the equator to high latitudes in the North Atlantic, giving southern England and western Ireland a much more temperate climate than they would otherwise have. There is also a worldwide current system at greater depth in the oceans which follows a different path to the surface water. It snakes its way through the world's oceans, driven by density contrasts in water. In the North Atlantic, near Iceland, the surface ocean water either evaporates in near-freezing conditions, or locally actually freezes itself.

Both these processes locally push up the salt concentration in the surface ocean water, which as a result becomes unusually dense. The dense water sinks right to the bottom of the ocean, flowing southwards down the length of the Atlantic towards Antarctica and meeting up with another deep flow of cold and salty water. The cold currents continue onwards, warming slightly and eventually rising to shallower levels in the northern part of the Indian and Pacific Oceans. This is often called thermohaline (temperature and salt controlled) circulation.

The global flow of cold and salty deep water sets in motion ocean currents at intermediate water depths. Where the cold and salty water first sinks to the ocean bottom, in the North Atlantic and Antarctic oceans, warmer water wells up from intermediate depths to the surface to replace it. This upwelling not only stirs up the nutrients and oxygen in the oceans, creating rich fishing grounds, but also warms the atmosphere, liberating an astonishing amount of heat. This is particularly important in the North Atlantic, where the upwelling water heats much of northern Europe, keeping it warmer than it would otherwise be.

Climatologists have identified a series of events at the end of the last glacial maximum which they believe highlight the importance of the thermohaline circulation in controlling the climate. The oxygen isotope record in ice cores from the centre of the Greenland ice sheet shows a steady warming after the last glacial maximum until about 13,000 years ago. Then, the climate swung back abruptly to much colder conditions. This is called the Younger Dryas event (see p. 154), marked by the spread of an Arctic plant *Dryas octopetala*. But, about 11,500 years ago, the average atmospheric temperature in Greenland began to increase quickly, rising by 7°C over the following fifty years. This would imply an average global change in temperature over the same period of about 4°C. If we place this in the context of present fears about global warming, which appears to taking place at a rate of about 0.1°C per decade, we can see that the temperature fluctuation about 11,500 years ago was truly gigantic!

The oceanographer Wallace Broeker believes that the Younger Dryas event was triggered by the stopping

## DEEP OCEAN CURRENTS IN THE ATLANTIC

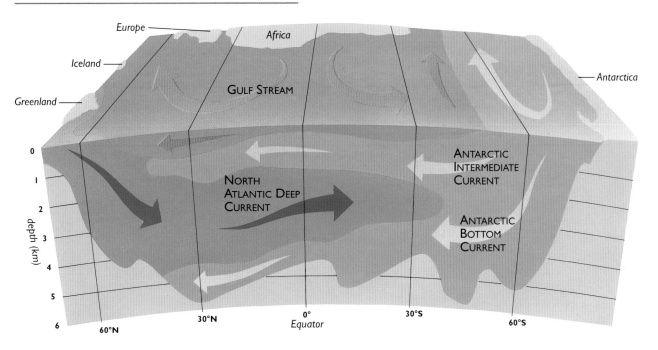

Water in the Atlantic Ocean is in constant motion. Surface currents, such as the Gulf Stream, move in great circular paths. But water at greater depths in the ocean has a different pattern of motion. The Antarctic Bottom Current and North Atlantic Deep Current consist of bodies of very cold, salty water which sink several kilometres in the polar regions and then flow either north or south down the length of the Atlantic. Compensating currents at shallower depths flow northwards. All these currents are called thermohaline circulation.

and starting of the thermohaline circulation. When the ice sheets melted at the end of the last glacial maximum, huge volumes of cold fresh water reached the sea. The idea is that about 13,000 years ago, the surface water in the North Atlantic became so diluted by this influx of fresh water that it was no longer salty enough to sink to the ocean bottom. This put a brake on the global thermohaline circulation, even to the extent of shutting it off. Northern Europe, deprived of the warming effects of the upwelling parts of the thermohaline circulation, was suddenly plunged into much colder conditions. Eventually, when the influx of melt water declined, the North Atlantic would have become salty enough again for the thermohaline circulation to suddenly start up again, triggering the dramatic warming at the end of the Younger Dryas event, 11,500 years ago.

If changes in the ocean current system can induce short-term fluctuations in the global climate, perhaps they can have a long-term effect as well? For instance, the marked global cooling, about 15 million years ago, could have been triggered when plate motions separated Antarctica and South America and the circum-Antarctic current started up. This current blocks the southward flow of warm water in the Pacific and Indian Oceans, keeping summer temperatures low in the Antarctic. Likewise, the closing of the Isthmus of Panama, as a consequence of plate motions in the Caribbean, shut off the equatorial connection between the Pacific and Atlantic Oceans, thereby changing the pattern of ocean currents. This might have been the culprit for the marked cooling, about 2.5 million years ago, at the start of the northern hemisphere glaciation.

There are yet other factors which can have a profound effect on the global climate. Dust particles from volcanic eruptions, swept into the upper atmosphere, can enhance global cooling by reflecting back more solar radiation into space. High mountain ranges can influence the workings of the atmosphere, by both deflecting the flow of air and warming higher levels of the troposphere. It also seems that the rise of

# The advance and retreat of ice sheets

*Glacial period – advance of ice sheets*

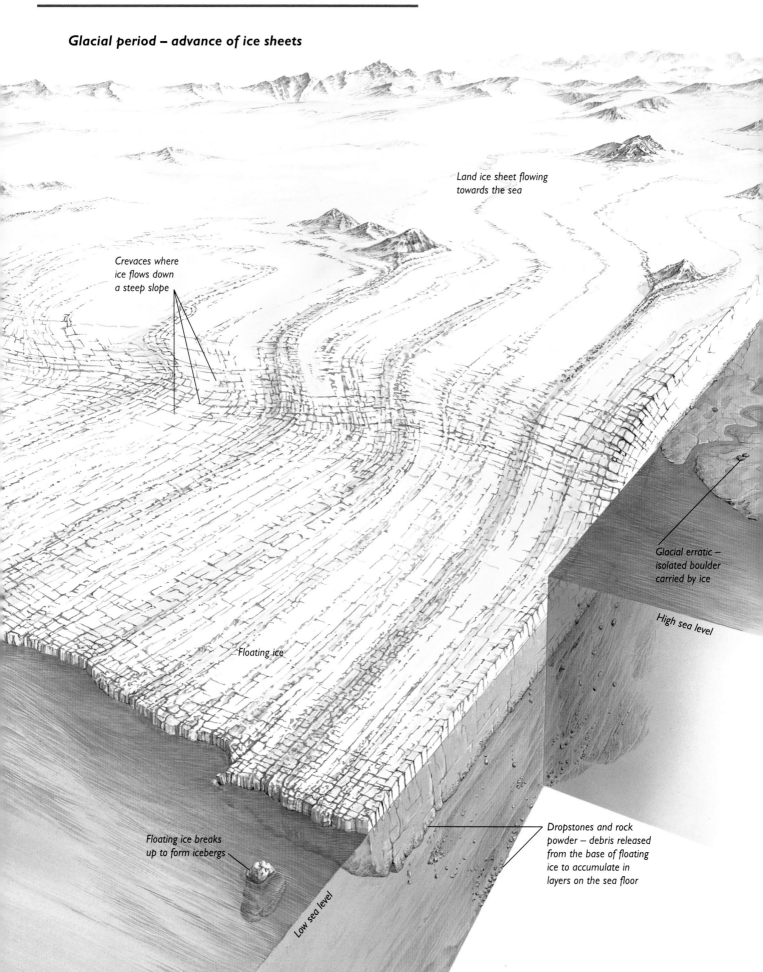

Land ice sheet flowing towards the sea

Crevaces where ice flows down a steep slope

Glacial erratic – isolated boulder carried by ice

High sea level

Floating ice

Floating ice breaks up to form icebergs

Low sea level

Dropstones and rock powder – debris released from the base of floating ice to accumulate in layers on the sea floor

## Interglacial period – retreat of glaciers

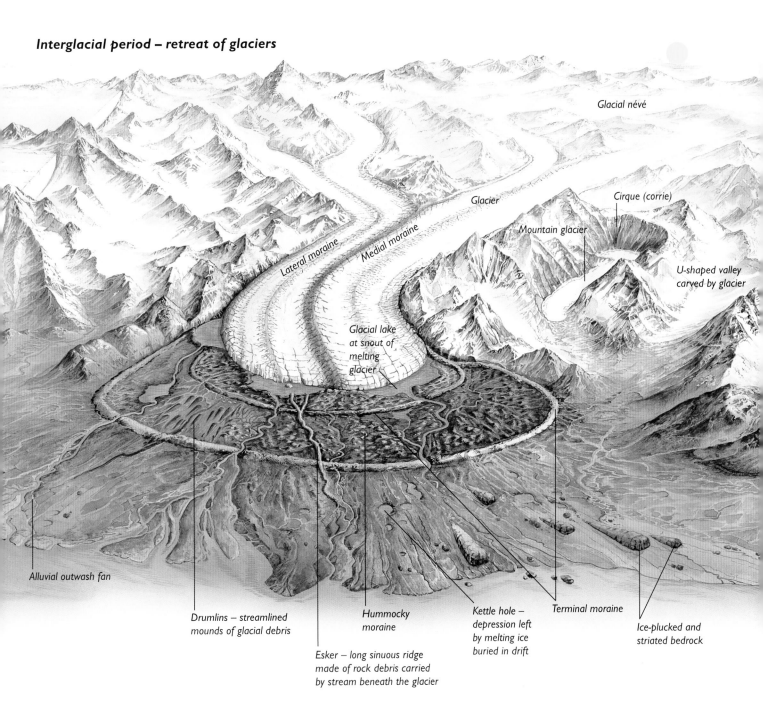

Glacial névé

Glacier

Cirque (corrie)

Mountain glacier

U-shaped valley carved by glacier

Lateral moraine

Medial moraine

Glacial lake at snout of melting glacier

Alluvial outwash fan

Drumlins – streamlined mounds of glacial debris

Hummocky moraine

Esker – long sinuous ridge made of rock debris carried by stream beneath the glacier

Kettle hole – depression left by melting ice buried in drift

Terminal moraine

Ice-plucked and striated bedrock

Over the past few million years the Earth has been in the grip of an Ice Age. There have been times, called glacial periods, when the climate was exceptionally cold and vast ice sheets, several kilometres thick, covered large parts of the continents (left). Sea level was lower than it is today. The immense weight of the ice caused the ice sheets to flow outwards, eventually reaching the sea. Rock fragments caught up at the base of the ice cut into the underlying bedrock, scraping up more rock debris which was eventually deposited beneath the ice sheet.

There have also been warmer times during the Ice Age, called interglacial periods, when sea level has been high (right). We are in an interglacial period today. These are periods when the ice sheets have retreated, and glaciers are mainly confined to mountainous regions. However, the traces of former ice sheets or large mountain glaciers can be seen in the striated and sculpted bedrock, U-shaped valleys, cirques, and mounds of chaotic rock debris called drift or moraine. Near the snout of a glacier, the drift often forms a distinctive landscape with hummocky and terminal moraines. Medial and lateral moraines lie within and to the side of the glacier.

the Tibetan plateau to a height over 5 kilometres caused a marked change in the pattern of the Indian monsoon about 8 million years ago (see Chapter 5). Plate tectonics may exert another intriguing influence. There is a general coincidence between Ice Ages and periods in the Earth's history when continents are clustered near the poles. For instance, during the early Ice Ages, around 600 million years ago, and also in a later period of Ice Ages, around 280 million years ago, large supercontinents were near the south pole. This may have had two effects on the climate. To begin with, land masses provide a surface for snow to accumulate on, and snow absorbs less of the incoming solar radiation than oceans, reflecting more of it straight back into space. Therefore, the presence of continents near the poles will tend to make the poles cooler than otherwise. Secondly, ice sheets on land can spread further. Today, the limit of the great Antarctic ice sheet is more or less the edge of the Antarctic continent, and floating ice extending beyond this is quickly broken up and dispersed. But in the northern hemisphere, during a glacial maximum, ice sheets spread southwards over land for thousands of kilometres.

## A CHANGEABLE WORLD

The Earth's climate has turned out to be extraordinarily complex, and scientists do not yet fully understand it. This should not surprise us because we are really looking at an intricate coupled system which links the behaviour of the atmosphere, oceans, land surface and the planet's orbit around the Sun. It has only been with the spectre of global warming looming on the horizon that scientists have started to come to grips with how this system works. They have had to employ the world's most powerful computers to do this, and even these are nothing like fast enough. But one lesson we have had to learn is that climate change is not an aberration or some malfunctioning of our planet; it is part and parcel of its natural behaviour. The Viking communities in Greenland, or ice skaters on the frozen Thames in the early part of the last century, were witnessing this natural variation. During the transition periods, when a glacial ends and the world begins to warm again, the landscape becomes alive with geological activity as glaciers and ice sheets melt and vast quantities of water flow over the surface of the continents. This powerful movement of water carves out much of the landscape of our world, as well as transporting large amounts of rock debris. The water pours into the oceans and sea level rises. Over longer periods of time, the rise and fall of sea level impose a fundamental rhythm on the pattern of sediment layers that accumulate on the margins of the continents.

Climatic change is something that the stream of life over the aeons has had to contend with. Indeed, it is the way the planet has had its most profound effect on the evolution of life. Slow changes, over tens of millions of years, provide a background evolutionary pressure. High frequency temperature fluctuations, over tens or thousands of years, abruptly weed out the less adaptable creatures. Something like the temperature changes at the end of the Younger Dryas event, nearly 11,000 years ago, are on a short enough timescale to pose a real threat to us. However, since then, during the Holocene period, the climate seems to have been unusually stable. Perhaps the evolution of human culture, which is closely connected with the development of agriculture, was possible only because of this brief spell of climatic stability. What is certain is that as in the past, there will be major changes in the Earth's climate in the future, irrespective of our own impact on the world.

*Left: When mountain glaciers reach a lake or the sea, the ice floats, sometimes breaking off to form icebergs – this glacier in the Patagonian Andes of Argentina is a good example.*

# CHAPTER 7

# THE LIVING PLANET

· · · · · · · · · · · · · · · · · · · · · · · · · · · · · · · ·

Over the past 4 billion years, life has evolved
from simple single-celled organisms into the
tremendous variety of plants and animals that exist
today. As scientists learn more about the Earth's
history, they are realizing that the forces
which have shaped the planet have also
had a profound effect on the course of evolution.
The movement of the tectonic plates has rearranged
the continents, providing ever-changing conditions for
living organisms, stimulating the evolution of new life-
forms. Violent volcanic eruptions, meteorite
impacts and drastic climatic changes have
triggered mass extinctions, causing setbacks
to life on Earth. But the same events have
provided new opportunities for the survivors.

*These cushion-like structures in Shark Bay, western Australia, are produced by algae
which thrive in the intertidal zone, building up the mounds, layer by layer. Fossilized examples,
called stromatolites, are some of the earliest evidence for life on Earth.*

The variety of living organisms is truly staggering, with millions of different species living in every conceivable corner of the planet, from superheated water in black smokers at the bottom of the deep oceans to the freezing conditions on giant ice sheets. In the last century, Charles Darwin dared to ask a question, which to many at the time seemed blasphemous: is there some fundamental mechanism which can explain this diversity? In 1859 he published his conclusions, in what turned out to be one of the most influential books ever written – *The Origin of Species*.

The first chapter of *The Origin of Species* is about domestic breeding. Darwin was brought up in a world of breeders: horse breeders, cattle breeders, pig breeders, sheep breeders, dog breeders. Darwin himself was a keen pigeon fancier and plant breeder. Talking to those other breeders, and from his own experience, Darwin became aware of what he termed the inherent 'plasticity' of life. Living creatures can be bred to virtually any size, shape or character. All one has to do is pick out from the progeny of a plant or animal those forms which have some similarity, however vague, to what one is after. If this process is repeated over many generations, a new breed, tailor-made by human hands for human purposes, is gradually created. It was not a large step from this observation to speculate that if humans can, by selecting for particular characteristics, guide the course of evolution of domestic animals or plants on a very short timescale, perhaps nature can do the same. Thus was born the seminal idea of natural selection. In Darwin's view of evolution, the competition between organisms drives evolution by constantly favouring any characteristic which gives the offspring a competitive edge. Gradually, over time, these 'selected' characteristics accumulate to result in an organism which is very different from its remote ancestors. This leads to a great diversity of life forms, especially as this process takes place in many different parts of the world.

Since Darwin first published his concept of natural selection, our understanding of biology has increased enormously. We now know that the characteristics of an organism are encoded in its genes, which are passed on to its offspring. The study of genetics has

*Animal breeders can create new varieties of organisms by carefully selecting offspring which have particular sought-after characteristics. Over many generations, these strange and exotic breeds of canary (centre and bottom) have been bred from the common canary (top).*

put a crucial part of Darwin's mechanism of evolution on a firm foundation. Nonetheless, in the past ten or fifteen years, as scientists have studied the fossil record, they have begun to suspect that there may be more to the evolution of life on Earth than originally envisaged in Darwin's theory.

## THE FOSSIL RECORD

Fossils have long been used by geologists to characterize particular sequences of rocks (see Chapter 1). But the existence of fossils is something of a geological miracle. After they die, most organisms quickly decay. But sometimes a sequence of events may lead to their preservation as a fossil. For example, organisms which die at the bottom of the sea or a lake may be entombed in soft mud or sand, becoming buried even deeper as more mud or sand accumulates above them. Organic matter decays very quickly as a result of bacterial activity; however, if the surrounding sediment is bathed in water, new minerals may slowly replace the decaying parts of the organism – for instance, bones or shell-like parts may be replaced by calcite or silica or iron sulphides. These chemical changes generally preserve the shape of the original parts of the creature. The degree of preservation can be truly remarkable, sometimes with the shape of individual cells faithfully preserved. Sometimes even the impression of soft parts of the creature, moulded in the surrounding sand or mud, may be preserved by the new minerals.

The study of the fossil record was pioneered in the last century. Geologists devoted considerable effort to collecting and painstakingly recording fossils. It rapidly became clear that most of the forms of life preserved in the fossil record do not exist today. But there were clear lineages of life through geological time – in fact, this is the main source of evidence for

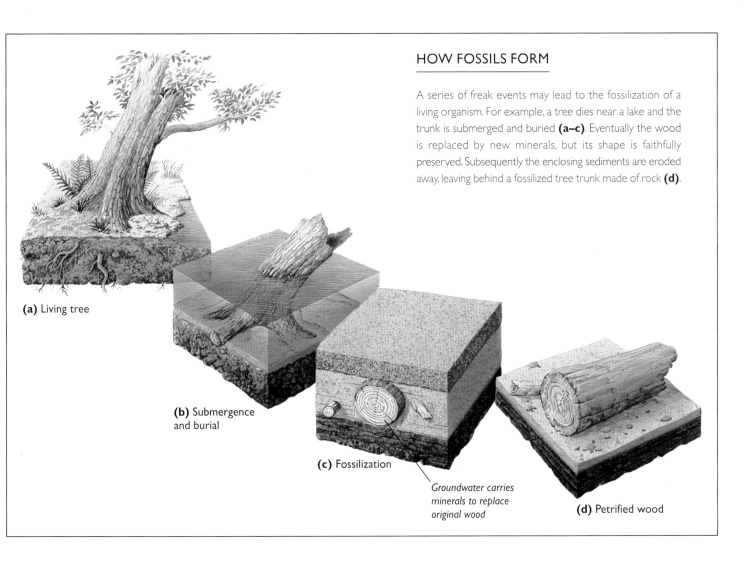

### HOW FOSSILS FORM

A series of freak events may lead to the fossilization of a living organism. For example, a tree dies near a lake and the trunk is submerged and buried **(a–c)**. Eventually the wood is replaced by new minerals, but its shape is faithfully preserved. Subsequently the enclosing sediments are eroded away, leaving behind a fossilized tree trunk made of rock **(d)**.

**(a)** Living tree

**(b)** Submergence and burial

**(c)** Fossilization

*Groundwater carries minerals to replace original wood*

**(d)** Petrified wood

the evolution of life. The fossils could be used to divide up geological time into distinct units. At first, time was divided into time before life and time with life. The latter, known as the Phanerozoic, was subdivided into three main periods called the Palaeozoic (ancient life), Mesozoic (middle life), and Cenozoic (recent life). These main periods could be further subdivided depending on the nature of the rock units or the various life-forms, giving rise to a series of units of geological time. Until the 1940s, there was virtually no evidence for life before the Palaeozoic, in a period referred to as the Precambrian. Today the fossil record can certainly be extended much further back in time, as we shall explain.

In the last twenty years, biologists have found a new way to study the evolution of life. They have started to decipher the code in the genes, in effect reading the 'blueprint' of a living organism. The code is, in fact, a sequence of molecules called nucleotide bases, which together make up DNA (deoxyribonucleic acid) – the genetic material in living organisms; different sequences result in different characteristics in the living organism. It is an extraordinary fact that the language of the genetic code is always the same, regardless of the organism; particular sequences of bases in the DNA virtually always have the same meaning. This is the strongest argument that life as we know it originated only once and that all living organisms are related, however remotely. Otherwise, one would expect the genetic code to differ, depending on which particular initiation of life organisms were descended from. Evolution at the level of the genes is a steady change in the sequence of bases in DNA. The comparison of these sequences from a number of different living organisms has helped scientists work out evolutionary relationships, pointing to gaps in the fossil record. This has made it possible to construct the tree of life, which shows the last common ancestors for different life forms, defining when different evolutionary trends branched off. The tree of life provides biologists with a unique way of classifying living organisms. Major branchings can be used to place organisms in a series of hierarchical categories. For instance, we classify ourselves as part of the animal *kingdom*, in the *phylum* of vertebrates (Chordata), class of *mammals* (Mammalia), *order* of apes (Primates), *family* of hominids (Hominidae), *genus* of man (Homo), *species* that is wise (sapiens). Kingdoms and phyla describe divisions that occurred near the roots of the tree of life, and living species are the tips of the youngest branches.

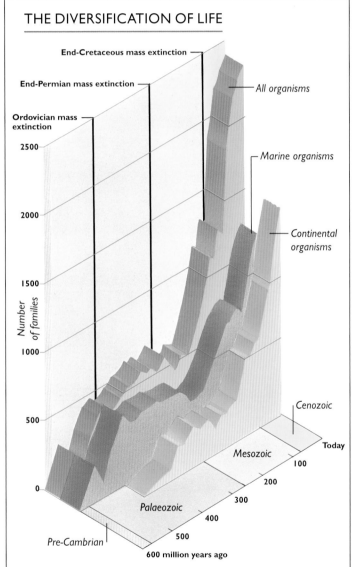

## THE DIVERSIFICATION OF LIFE

Living organisms can be classified into different groupings such as families. The fossil record shows that during last 600 million years living organisms have diversified, so that the number of families of both marine and continental organisms has increased markedly from a very few. However, the rate of increase has varied through time and there have been a series of temporary setbacks, called mass extinctions, when the number of familes has declined abruptly before increasing again. Notable extinction events occurred at the end of the Ordovician, and at the ends of both the Permian and Cretaceous periods.

Looking at the fossil record in the context of the tree of life has provided some intriguing insights into the nature of evolution. Firstly, it is striking that most branches come to a dead end. In fact, almost all the species that ever lived have subsequently died out. Despite this, the number of different species, genera and families living on Earth at any time has increased markedly over the course of evolution. Life is probably more diverse now than it has ever been. But this increase in diversity has been far from steady. Rather, there appear to be times in Earth history when there was a sudden increase in the number of different life-forms, and other times when the variety of living things suddenly decreased. Sometimes these setbacks to diversity have been the result of a wholesale slaughter of life – a mass extinction. Each mass extinction has proved to be a turning point in the development of life as the extinct species are replaced by an even greater variety of new creatures, often characterized by the sudden appearance of novel features. Thus the slow and stately progress of evolution, envisaged by Darwin, seems to have been interrupted by a number of abrupt events that have had a decisive impact on the direction life has taken. Geologists are now beginning to link these dramatic episodes in evolution to major changes in the Earth. In doing so, they are revealing unexpected powerful links between our planet's geological activity and the evolution of life. Our world does not merely provide living organisms, as it were, with an address – it has been an active player in the drama of evolution, often dictating the course of events which has resulted in the life around us.

We can explore these links by following the story of life from its origin to the present day. We rely on a number of key pieces of evidence to put together this story. Firstly, the fossil record, combined with studies of the genetic material of living organisms, reveals the main features of the tree of life. The rock record tells us about the conditions on Earth at each stage in the evolution of life. We can track the motions of the continents across the surface of the globe (see Chapter 2). When continental fragments break apart or collide, they leave their mark in the rock record. Dating the rocks gives us the timescale for this story.

## LIFE GETS GOING

The earliest evidence for life found so far is in a 3.8-billion-year-old banded sequence of silica and iron-rich rocks called Banded Iron Formation, or BIF for short, from western Greenland. These rocks are commonly found in ancient greenstone belts (see Chapter 1). BIFs consist of millimetre- to centimetre-thick layers of iron oxides, such as haematite and magnetite, alternating with silica-rich layers, which are thought to have formed by precipitation from sea water. The evidence for life in them does not come from fossilized remains, but from a peculiar chemical signature of living organisms. This is the ratio of two natural isotopes of carbon – carbon-13 and carbon-12. Living organisms tend to extract more carbon-12 from the environment, at the expense of carbon-13. Thus, the ratio of the number of atoms of carbon-13 and carbon-12 is typically less in living organisms than elsewhere on Earth. Traces of carbon have been found in the early Greenland BIFs and these have a carbon isotope ratio similar to that expected for living organisms. This carbon may be all that remains of organisms that lived roughly 3.8 billion years ago – traces of the oldest life-forms found so far on Earth. In this case, life was under way within a maximum of 700 million years of the formation of the Earth, now dated to 4.55 billion years ago – less than a sixth of the way through its history.

In 1996 Maarten de Wit, who has worked extensively on some of the oldest rock formations (see Chapter 1), made a remarkable discovery, together with the palaeobiologist Frances Westall, in the Barberton Greenstone Belt, southern Africa. They found the actual fossilized remains of the oldest-known life forms on Earth: rice-grain shapes a few thousandths of a millimetre across, perfectly preserved in rocks called cherts. These have the identical shape to that of bacteria living today. The cherts formed when silica was precipitated from hot springs in a region of intense volcanic activity. Evidence for this is found in the adjacent rocks – huge volumes of

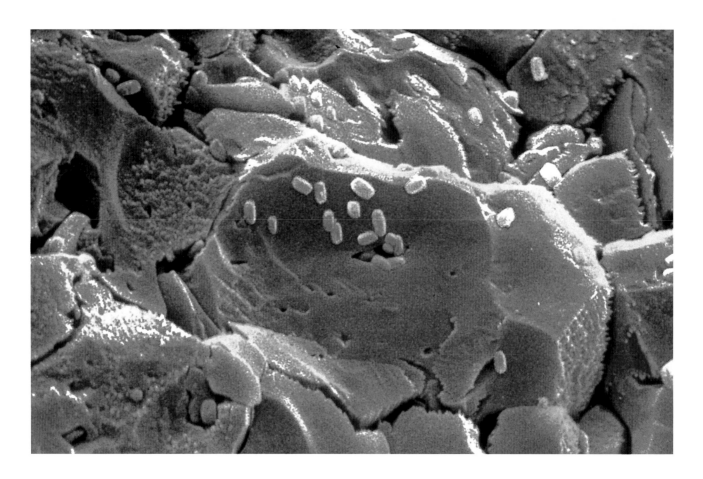

Rod-shaped bodies (coloured red here), identical to bacteria, are revealed in 3.5 billion-year-old silica-rich rocks from South Africa when viewed with the scanning electron microscope (field of view is about 3 microns across). Frances Westall, who found them, believes that they are the oldest fossils known.

pillow-shaped lavas which erupted under water. Dating the lavas shows that the fossilized cells were thriving 3.5 billion years ago. The rocks give us a picture of life on the early Earth, full of heat-loving microscopic organisms living in numerous hot springs, powered by the intense volcanic activity of the early Earth.

Bacteria are the most prolific example of a type of single-celled organism called a prokaryote. So, what were the steps that led from no life to the evolution of a prokaryotic cell? These steps must result in all the crucial characteristics of the cell: the presence of genetic material in the form of DNA, a semi-permeable membrane which allows the cell to selectively absorb chemicals from the outside world, and a mode of organic chemistry which harnesses a flow of energy. The close association of the primitive

prokaryotes with hot springs suggests that life may have started in such an environment, exploiting the flow of thermal energy from the Earth's interior. Further evidence for this has been found at the mid-ocean ridges – we have already described some evidence of this in Chapter 2. Here, prokaryotes which have proved to be extremely primitive – rooted in one of the lowest branches of the tree of life – cluster around vents of superheated water called black smokers. They are capable of withstanding water temperatures up to 100°C, and for this reason are

Right: Geysers, like this one in Yellowstone Park, USA, may have been places where the earliest forms of life evolved. Some of the most primitive organisms found today are bacteria which thrive in the hot water, not only on land, but in deep sea hot springs as well.

*Crumpled red layers (top half of picture) are the remains of vast deposits of iron oxide, called Banded Iron Formation, which were laid down about 2.5 billion years ago in what is today the Hamersley region of western Australia. These could only have formed if the Earth's atmosphere contained virtually no oxygen.*

called hyperthermophiles. The hyperthermophiles exploit the energy released when the chemical-rich hot water mixes with cold sea water in a process called chemosynthesis. A whole cocktail of molecules which are commonly found in living organisms is produced this way, including simple amino acids which are essential for many of the molecules of life.

It is possible to imagine a situation where a stable chemical environment was set up within the confines of an inorganic semi-permeable membrane. One

suggestion is that primitive semi-permeable membranes were cell-like silica or metal crusts. Such natural bubbles are commonly observed today around hot springs. The hot water in such a volcanic environment would be rich in many chemicals, including metal ions, essential for life-like reactions. Nucleotides and other molecules could come together to form simple macromolecules – this way we would have the beginnings of genes. If by chance any of these macromolecules were capable of replication, then its concentration would start to increase. At some point, the original inorganic membranes may have become templates for organic membrane constructions; at that point the membranes and their contents could have become free cells. If by chance one of these membranes got pinched off, like a soap bubble, the cell would have effectively reproduced. Any membrane system that had a tendency to split would increase in number. And so on. The steps are all very speculative, but in principle it is possible to conceive how, through natural selection, a replicating single cell could come about.

If the ideas discussed above are correct, then life at its earliest stages was, like a newly born baby, dependent on its mother, the Earth, deriving both sustenance and energy from the planet's volcanic activity. However, there was another huge source of energy: the rays from the Sun. When living organisms started using this is unknown, though there is a suggestion that the ability to detect hot water – an evolutionary adaptation which would have increased the chances of survival for a prokaryote dependent on hot water at just the right temperature – evolved into the ability to harness sunlight. Certainly, once the prokaryotes were weaned off the hot springs, and were living on sunlight in a process called photosynthesis, they could break free from their volcanic prisons and colonize the Earth as a whole. And living organisms seem to have done this with a vengeance, even to the extent of influencing, as we shall see, the planet itself.

Photosynthesis uses sunlight to convert carbon dioxide and water to organic matter and oxygen gas, and is, today, one of the most important chemical reactions in the living world. This is the first step in building and powering plants upon which nearly all living organisms ultimately rely. Photosynthesis has also turned out to be the main source of free oxygen gas on the Earth. Today, photosynthesizing cyano-bacteria – a type of algae – are uniquely responsible for a distinctive structure called a stromatolite, which is found in a few places on Earth. Shark Bay in western Australia is one example. Here, mats of photosynthesizing algae trap particles suspended in coastal waters. The metabolic activity of the algae seems to promote the precipitation of calcium carbonate from sea water, coating the algae and trapped material. The algae grows through this coating, trapping more particles, and the whole process is repeated. Eventually, a mound-like structure builds up, containing many fine-scale layers. Stromatolites have been found in rocks as old as 3.5 billion years, strongly suggesting that photosynthesizing algae also existed at this time. By 2.5 billion years ago, there were substantial continental areas where algal mats could thrive, building up to form wide areas crowded with stromatolitic mounds.

It seems to have taken well over a billion years before the photosynthetic activity of algae had much effect on the atmosphere. We know this because huge BIF deposits, covering an area of thousands of square kilometres, were laid down in what is today the Hamersley region of western Australia about 2.5 billion years ago. Much of the iron in these deposits appears to have been washed into the region, carried by a major system of rivers. For the iron to be deposited in concentrated form, it must have been carried in solution by the rivers. If the atmosphere was rich in oxygen at this time, the iron would soon have oxidized, precipitating out in the soil and rivers, never reaching the shallow sea where the Hamersley BIFs formed. So the level of oxygen in the atmosphere must have been low.

## LIFE GETS COMPLEX

About 2.1 billion years ago, the concentration of oxygen in the atmosphere began to increase markedly. The evidence for this is in the rocks. Extensive banded iron deposits (BIFs) ceased to form in lakes and the sea. Conglomerates full of grains of pyrite (iron sulphide) and uranium minerals, which are unstable in the presence of oxygen, no longer accumulated. It seems that the oxygen produced by photosynthesizing prokaryotes exceeded, for the first time, the amount of oxygen that was taken out of the

# The evolution of life

This diagram charts both the evolution of life and major events in the history of the Earth. Over the past 4 billion years life has evolved from microscopic single-celled organisms to the present diversity of living creatures. But our planet has not merely provided living organisms, as it were, with an address, it has been an active player in the drama of evolution. At various times in the past, the geological activity of the Earth, manifested in the drifting

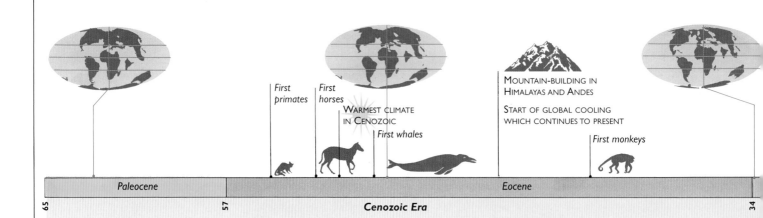

First primates

First horses

WARMEST CLIMATE IN CENOZOIC

First whales

MOUNTAIN-BUILDING IN HIMALAYAS AND ANDES

START OF GLOBAL COOLING WHICH CONTINUES TO PRESENT

First monkeys

| Paleocene | Eocene |
|---|---|

65  57  **Cenozoic Era**  34

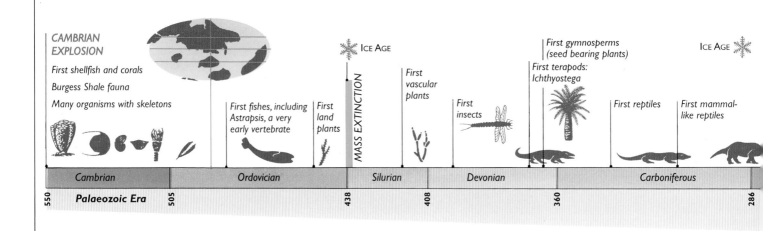

CAMBRIAN EXPLOSION

First shellfish and corals

Burgess Shale fauna

Many organisms with skeletons

ICE AGE

First vascular plants

First gymnosperms (seed bearing plants)

First terapods: Ichthyostega

ICE AGE

First fishes, including Astrapsis, a very early vertebrate

First land plants

MASS EXTINCTION

First insects

First reptiles

First mammal-like reptiles

| Cambrian | Ordovician | Silurian | Devonian | Carboniferous |
|---|---|---|---|---|

550  **Palaeozoic Era**  505  438  408  360  286

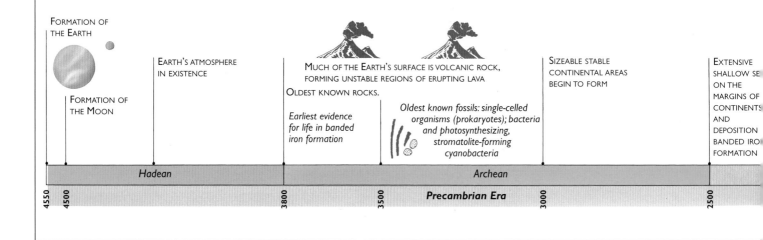

FORMATION OF THE EARTH

EARTH'S ATMOSPHERE IN EXISTENCE

MUCH OF THE EARTH'S SURFACE IS VOLCANIC ROCK, FORMING UNSTABLE REGIONS OF ERUPTING LAVA

OLDEST KNOWN ROCKS.

SIZEABLE STABLE CONTINENTAL AREAS BEGIN TO FORM

EXTENSIVE SHALLOW SE ON THE MARGINS OF CONTINENTS AND DEPOSITION BANDED IRO FORMATION

FORMATION OF THE MOON

Earliest evidence for life in banded iron formation

Oldest known fossils: single-celled organisms (prokaryotes); bacteria and photosynthesizing, stromatolite-forming cyanobacteria

| Hadean | Archean |
|---|---|

4550  4500  3800  3500  **Precambrian Era**  3000  2500

of the continents, or periods of intense volcanic activity, or even meteorite impacts, has had a profound effect on the evolution of life. These events precipitated drastic changes in the climate, such as the initiation of Ice Ages, which ultimately caused mass extinctions of life. Only those organisms which were versatile enough to survive these periodic crises could carry forward the evolution of life.

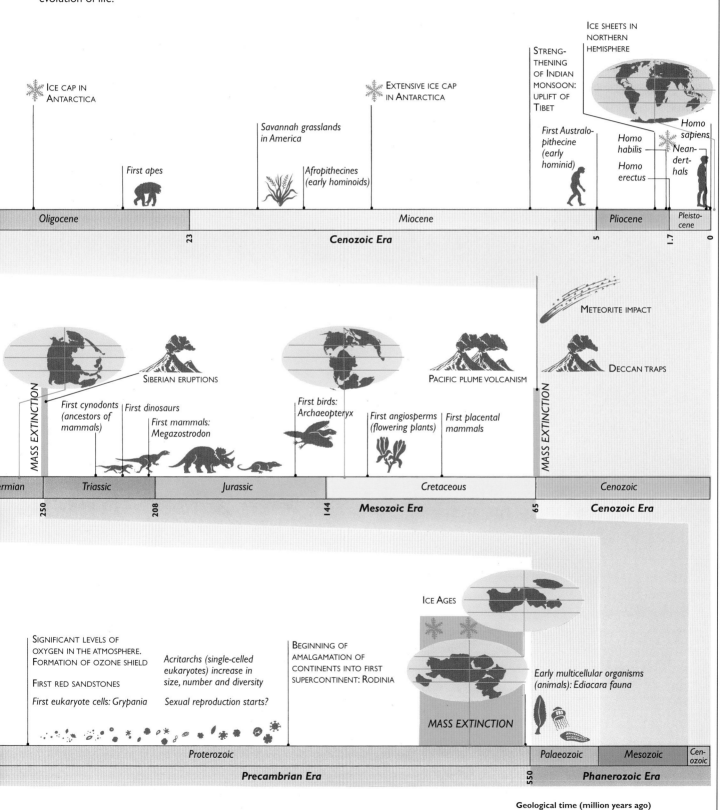

Geological time (million years ago)

oceans by oxygen-consuming reactions such as the oxidation of soluble iron, sulphur and dead organisms. This may have been partly because the supply of soluble iron to the oceans from volcanic sources had also decreased, perhaps as volcanism waned with the slow cooling of the Earth. Oxygen in sufficient quantities could now get into the atmosphere, so that iron rusted on land, staining the rocks red. This way, distinctive deposits of iron-stained sandstones, called redbeds, began to appear in the rock record. It is likely that the atmosphere about 2.1 billion years ago was sufficiently rich in oxygen for the creation of ozone in the upper atmosphere to be an important process. This way, the Earth gained an ozone shield. Ozone is made up of three linked oxygen atoms, created by the dissociation of normal oxygen gas in sunlight. Today, the ozone layer in the upper atmosphere protects us by absorbing much of the harmful radiation from space, such as certain bands of ultra-violet light (UV-b). The living descendants of the early prokaryotes are the hardiest organisms on Earth, surviving in virtually any environment. This hardiness may reflect the characteristics early prokaryotes needed to survive on the Earth before it had its ozone shield.

The presence of oxygen in the atmosphere had several profound effects on the evolution of life. For the first time, it would have been a positive advantage for living organisms to make use of the abundant energy released when oxygen reacts with organic matter – the process of oxidation. This reaction is potentially lethal for a cell, as it can cause the very substance of the organism literally to burn. But a cell which by chance developed the capability to handle this gas and exploit the energy released when it reacts would multiply rapidly. Because of the ozone shield, it was easier for complex organic structures to exist in very shallow water or on land and make use of the solar radiation without being fried. At first, a number of different prokaryote cells may have lived together, as it were, under one roof in a larger cell. Eventually, the genetic material from each of these was pooled together in a special compartment – perhaps a defunct prokaryote – to form a cell nucleus with a large amount of genetic material. Individual members of this prokaryotic 'community' may have started to specialize, taking on particular functions of cellular activity; this way compartments within the cell with distinct functions, called organelles, may have come into being. Thus was born the 'oxygen-breathing' eukaryotic cell, with its more complex internal structure and nucleus, marking a major milestone in the evolution of life. More genetic material allowed more characteristics to be encoded in the genes, and also increased both the chance of a mistake in the copying process during cell division, and the chance of a mutation from the effects of radiation. And so the inherent variability of eukaryotic cells became substantially greater than that of prokaryotes – increasing the repertoire of characteristics available for natural selection when the environment changed and opening more evolutionary possibilities. Life started to become more varied.

Just when eukaryotes evolved is unclear. The cell nucleus and organelles are not preserved during fossilization, so whether a fossil cell is a eukaryote or prokaryote is largely determined by its size. Eukaryotes are much larger than prokaryotes. A spaghetti-like fossil called *Grypania*, which is substantially larger than normal prokaryote cells, has recently been found in a 2.1-billion-year-old iron formation in Michigan, North America, suggesting that eukaryotes started to appear very close to the time when the oxygen levels in the atmosphere increased and the ozone layer started to form. Large single-celled eukaryotes, called acritarchs, become more and more common and varied in progressively younger rocks, until about 900 million years ago when a huge number existed, some with cells up to 75-thousandths of a millimetre wide. These acritarchs were over a hundred times larger than the prokaryotes from which they originally evolved.

Between roughly 900 and 600 million years ago, the acritarchs had to contend with a major change in the global climate, and the fossil record suggests that they did not fare well. The decline of the acritarchs has been described as the first great mass extinction on Earth. It coincides with a series of Ice Ages in perhaps the coldest period in the history of the Earth. During this time, the continents had come together to form the first supercontinent called Rodinia. Evidence for the early stages in the amalgamation and rearrangement of the continents to form Rodinia are found in the eroded roots of ancient mountain belts in North America and Scandinavia. These are the remains of a period of continental collision between 1.3 and 1.1 billion years ago; 750 million years ago, Rodinia straddled the equator. But the remains of glacial gravels and glacial scratchmarks on rock surfaces found today in Africa, Australia, South America, northern Europe and North America, suggest that as Rodinia moved towards the south pole, ice sheets started to build up. The Earth was now in the grip of one of the major Ice Ages, called the Verangerian, and the ice sheets extended towards the equator.

## THE RISE OF ANIMALS

The final phase of the global cooling in the Precambrian world seems to have occurred about 590 million years ago. When the planet started to warm again, a new world of life was established: multi-cellular animals. In 1946, some extraordinary fossils, up to several centimetres long, were found in the Precambrian (Vendian) rocks of the Flinders Ranges in South Australia. Subsequently, examples have been found in similarly aged rocks all over the world, including those from Charnwood Forest in Leicestershire. All these fossils are now called the Ediacara fauna (i.e. animal-like creatures) after the mine where the first fossils were collected. The fossils consist of the impressions of soft-bodied, elongate leaf-shaped and round jellyfish-shaped creatures, preserved in sandstones and siltstones. Some scientists have argued that a number of the creatures in the Ediacara fauna are closely related to living jellyfish. Others believe that they are a strange aborted branch of the tree of life which, through a geological miracle, have been exceptionally well preserved. But their existence shows that complex life-forms were evolving at this time, with some of the outward appearances of animals.

Around 550 million years ago, the tree of life started to show an amazing spurt of growth, branching dramatically in many different directions. This is sometimes described as the Cambrian explosion of life, and marks the beginning of the 'time of life' (Phanerozoic), first recognized by nineteenth-century geologists. During this period, the number of orders of animals doubled roughly every 12 million years, showing an exponential increase, and almost all the present-day phyla of animals started to appear in the fossil record. Modern representatives of these phyla include worms, crabs, shellfish, lamp shells, sea urchins, sponges, tunicates and centipedes. Many of these creatures have an important feature which is more easily preserved in the fossil record: a skeleton, often forming spines, scaly plates, walls or shells. This skeleton was created by the precipitation of certain

*Impressions of strange jelly-fish-like organisms, such as this example from Australia, are found in rocks about 550 to 600 million years old. These fossils are collectively called the Ediacara fauna and may be the remains of some of the earliest multicellular animals. They probably lived on the sea floor at depths of about 100 metres.*

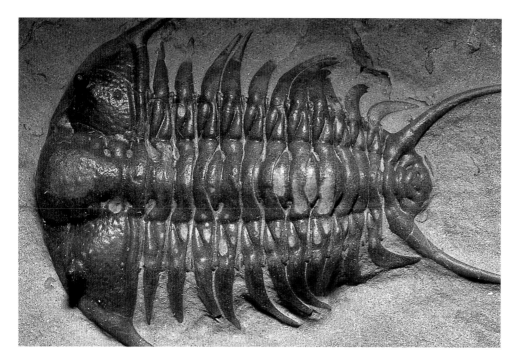

*Left: Trilobites are a ubiquitous fossil in marine rocks from the Palaeozoic ('time of old life') period. This example lived about 475 million years ago, and was a bit like a giant woodlouse, several centimetres long, crawling around on the sandy bottom of the sea.*

*Below: A rock formation in the Canadian Rockies, called the Burgess Shale, has proved to be a window on life over 500 million years ago in the Cambrian period. The fossilized remains of diverse and strange-looking creatures, like this arthropod, a few centimetres long, may be evidence for the first living communities where predator and prey alike had to survive alongside each other.*

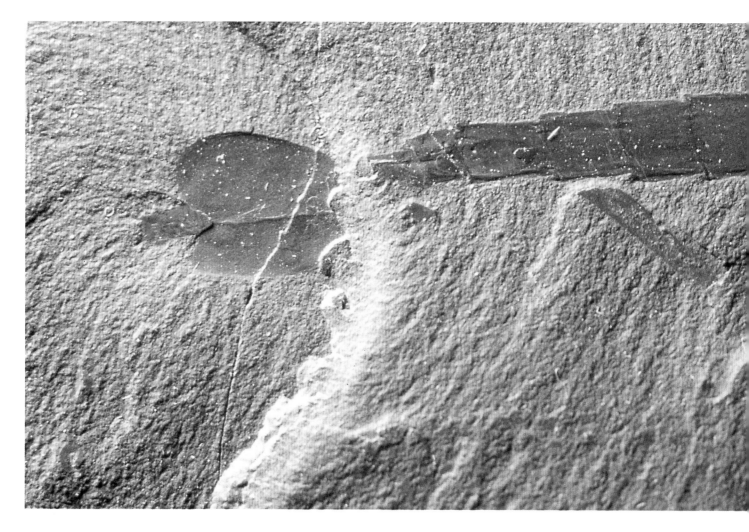

minerals – silica, calcite or calcium phosphate – controlled by the metabolic activity of the organism.

The suddenness of the Cambrian explosion demands an explanation, and geologists have considered many factors. It is possible that life was spurred on by the break-up of the Rodinia supercontinent, perhaps because this created just the right distribution of oceans and continental masses for a major ocean current system to be set up, stirring the ocean waters and distributing oxygen and other nutrients. It also seems that continental drift was exceptionally rapid at this time, reaching speeds of 20 centimetres per year. As the continents moved, ocean basins would have opened up rapidly, and the new and hot mid-ocean ridges would have been high enough to precipitate a rise in global sea level. The rise in sea level flooded many coastal regions, creating huge near-shore shallow and deep water environments,

covering large parts of the continents. Perhaps this provided the right sort of environment for the new phyla to exploit. Also, there may have been changes in sea water chemistry, as sea water was flushed through the ocean crust at the hyperactive mid-ocean ridges, and rock fragments, with their own chemical signature, were eroded from the mountains and dumped in the oceans. This may have promoted the development of a skeleton or hard parts in living organisms.

None of the previous explanations is totally satisfactory. This may be because scientists are looking in the wrong place. The Cambrian explosion may just have been the inevitable consequence of the existence of multicellular organisms, allowing ever more diverse and complex organisms to evolve. In other words, the origin of the subsequent explosion lies in that time just after the Verangerian Ice Age,

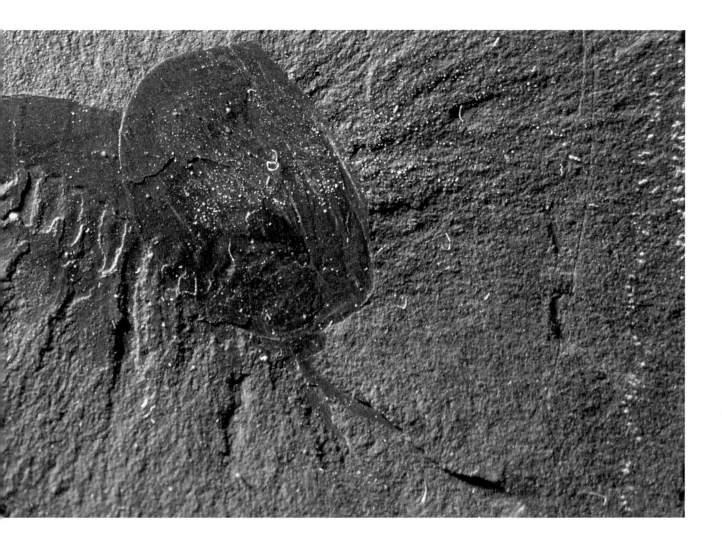

*The shallow sea floor, about 470 million years ago, in the Ordovician period, was populated by many creatures with hard skeletal or shelly parts, such as these squid-like cephalopods, brachiopods, trilobites and solitary corals.*

when the Ediacara fauna first appeared on the scene. As life diversified, the first living communities came into being, in which organisms interacted with each other. The complex webs of living organisms would be vulnerable to collapse if there was a change in the environment. Fossils showing signs of boring and other damage suggest that all was not harmonious in the communities and violent predators were out and about.

The decline in the growth of stromatolitic structures at the end of the Precambrian also suggests the presence of 'grazers' which attacked the stromatolite-building algal mats. In this case, an important agent of natural selection, driving evolution during the Cambrian explosion, was the competition between organisms. The early skeletons may have evolved as a

form of armour, protecting the organism from attack. If the selective pressure was sufficiently intense, a great diversity of animals with hard parts could emerge in a relatively short time in a sort of biological arms race. Further evidence that this might be at least a partial explanation for the Cambrian explosion comes from an extraordinary variety of fossils, called the Burgess Shale fauna, found in Cambrian black shales in British Columbia, Canada. Many of these creatures had spines, armour plates and grasping arms, suggesting an environment where interaction between Cambrian animals was much less benign than had been previously thought.

The evolution of a skeleton, whether as a direct consequence of predation or changes in the environment, had one very important consequence.

It provided a scaffolding on which a multicellular organism could grow. Without such a scaffolding, there would be severe limits to the size that life-forms could attain without becoming unwieldy. Within its armour, both the number and type of cells of a multicellular organism could increase without necessarily making the organism uncompetitive. In fact, quite the reverse was possible. The evolution of multicellular organisms with internal cell specialization allowed adaptation to ever more restricted environments, opening up whole new avenues for life. The great diversification of organisms continued, though interrupted by occasional setbacks as some taxonomic groupings inevitably became extinct, until about 440 million years ago at the end of the Ordovician period. The Ordovician radiation of life tripled the existing diversity. It may have been aided by the distribution of the continents at that time, spread out near the equator with large areas of warm water covering the continental shelf regions. Gradually, over the Ordovician period, living communities established themselves across the full width of these shelves. Silvery fish-like creatures also lurked in the Ordovician seas – one such is called *Astrapsis*.

At the end of the Ordovician period, the continents had rearranged themselves and a large landmass, comprising most of the present-day southern continents, drifted away from the equator towards the south pole. This seems to have triggered global cooling and the Earth was plunged into another Ice Age. Evidence for this can be found in the distinctive glacial gravels preserved in North Africa, Brazil and Arabia. By about 440 million years ago, the region which today is North Africa lay over the south pole. As the ice sheets expanded, sea level dropped. The shallowest parts of the sea were drained, killing the diverse living communities which had colonized these regions. Oscillations during the Ice Age, as ice sheets advanced and retreated, caused oscillations in sea level and rapid environmental change, further stressing and destroying living communities, and warm water coral reefs died. These extinctions mark the second great crisis in the evolution of life, when nearly three-quarters of all ocean species were wiped out in a period of less than a million years.

## KEEPING UP WITH THE CLIMATE

Already, at the dawn of the era of animals, in the latest Precambrian and during the Palaeozoic, a recurring theme in the evolution of life is clear. Living organisms diversify, expanding into new environments until they are checked by global climate change – often one of the Earth's periodic Ice Ages. Many life-forms are killed off, resulting in mass extinction. The survivors soon bounce back to recolonize the planet and evolve. They are clearly the ones who can cope better with environmental change, be it cooling, a drop in sea level, or changes in the oxygen levels of the oceans. In effect, climatic changes have weeded out the less adaptable life-forms, acting as an agent of natural selection. This is how the planet exerts its main influence on the evolution of life. Variations on this theme are repeated many more times in the Palaeozoic era. Indeed, this pattern continues for the rest of the story of life, though new agents of climate change appear on the scene – gigantic volcanic eruptions and meteorite impacts.

Another striking feature of the extinction and recovery pattern of life is that after many mass extinctions, some of the survivors evolve into ever larger life-forms. Thus the earliest life consisted of tiny single cells. At the end of the Precambrian, the largest creatures were only a few centimetres long. In the Cambrian and Ordovician, creatures tens of centimetres long existed. Subsequently, veritable monsters of the deep evolved, reaching several metres in length.

Plants show a similar pattern. The earliest land plants in the Ordovician were only a few millimetres high. But gradually, giants of the forest evolved, towering tens of metres above the ground. The plant or animal monsters multiply until the next major phase of climate change, when they are vulnerable and often become extinct. But, almost literally from under their feet, the midget competition which had managed to survive emerges to thrive and diversify, expanding both physically and in numbers.

Here, we might be witnessing the fundamental tension between the two agents of natural selection which drive evolution: competition and changes in the physical environment. Perhaps competition dominates in the long periods between severe environmental change and, in a predatory world, often drives organisms to evolve to ever larger sizes. Rapid climate change breaks the vicious cycle of competition, which drives evolution towards greater specialization in ever more restricted places – niches – of the environment. In these circumstances, trying to move from one niche to another is usually fatal; a better-adapted organism is sure to exist, and so natural selection favours specialists which have evolved to exploit their existing niches to the full. After severe climate change, many niches may be emptied of living organisms, and natural selection will instead favour any organism which is sufficiently adaptable to move into a new niche. Thus rapid climate change selects for more than robust and specialized living machines; it also favours versatile behaviour and a larger nervous system. For only organisms with these characteristics will emerge through the climate change hurdle, which the planet, from time to time, throws up in the way of the stream of life.

## LIFE LEAVES THE OCEANS

Throughout the Palaeozoic, or time of 'old life', the continents were slowly reassembling themselves, after the break-up of Rodinia, into a new supercontinent called Pangea. Already, at the end of the Ordovician, when so many organisms went extinct, the southern continents were assembled as Gondwanaland. For the rest of the Palaeozoic, the northern continents would slowly converge on Gondwanaland. During this period, the major event in the evolution of life was the transition of animals from water to dry land. It seems that a particular feature of the planet made this possible: the gravitational pull of the Moon on the Earth. This pull tends to bunch up the oceans, creating the twice-daily high and low tides as the

By 380 million years ago, in the Devonian period, forests of large trees had colonized the land. These fossilized logs from Arizona, USA, are the remains of trees which lived about 230 million years ago in the Triassic period.

Right: Vast swamps existed on the margins of the continents in the Carboniferous period, about 320 million years ago. They were probably very much like this view of the Everglades in Florida, USA. Dead plants, instead of decaying, sank into the mud and eventually became coal.

Earth rotates. The intertidal zone is a sort of halfway house between land and sea, and provided the necessary bridgehead for aquatic life on dry land. Land plants probably evolved via the descendants of various forms of seaweed and other algae, which thrived in the intertidal zone and were gradually able to tolerate ever longer exposure to the air, moving up the beach to the extreme high tide zone and then eventually out of the tidal zone altogether.

In the Ordovician, early forms of liverwort-like plants were already beginning to colonize boggy regions and ponds on land. In the ensuing Silurian period, land plants began to take a hold on the continents, developing from simple padded surfaces to vascular plants with stems and leaves, capable of regulating water loss through special pores called stomata. The land offered a whole new environment which at first could be colonized without competition. It also provided a stable base, unaffected by the trials and tribulations of water action and predatory marine animals. Plants could benefit from direct sunlight for photosynthesis, rather than having to make use of the filtered rays under water. The fossil record shows that Late Devonian forests, spread out along river valleys, contained trees over 10 metres high. The land was becoming a biological factory, making use of the abundant light energy to split carbon dioxide and water into the basic building blocks of life. This process was locking up large quantities of chemical energy – an energy source which could be harnessed by large animals if only they could get hold of it! For this to happen, two fundamental difficulties had to be overcome. Firstly, organisms have to cope with oxygen as a free gas in the atmosphere, and in much higher concentrations than in water. Secondly, out of the water, living organisms are exposed to the full force of gravity and must carry their whole weight if they are to move.

By the Early Devonian, worms, snails, myriapods (ancestors to centipedes and millipedes) and scorpions had arrived on dry land. Insects sub-

*Scaly fish, like this one (Thursius pholidotus), beautifully preserved in Devonian sandstones from Scotland, were living in lakes about 380 million years ago.*

sequently evolved from the myriapods. But other animals had to wait for an important development in the evolution of fishes before they could make it on to dry land. This was the evolution of an internal mineralized skeleton. The early fish fossils show no evidence for actual bones – the hard parts seem to have consisted of scaly plates and spines. The internal support is likely to have been an internal stiffening rod called a notochord, which was sufficient in the effectively near-weightless conditions under water. For reasons that are not entirely clear, this started to be mineralized, leading to the creation of fish bones. Thus the backbone of some Devonian fish became a series of vertebrae held together by tissue and supporting the rest of the body. These fish were the earliest vertebrates and had evolved a basic structure which has not changed for the rest of the evolution of animals: a bony skeleton with spine and limbs.

The bony vertebrate structure that already existed in some fish was the prototype of the weight-supporting skeleton which enables land animals to lift their bodies off the ground and move around. The very latest Devonian fossil record contains the first evidence for a true amphibian – *Ichthyostega*. This essentially looked like a fish with legs, walking around on the land with a marked sideways swaying motion. This early amphibian must have had to return to the water to reproduce. The evolution of the amniotic egg made permanent habitation on land possible. This egg contains a micro watery environment in which the young can go through the early stages of development as though in water. They can hatch only when their development has progressed to a point where they have the requisite characteristics for survival on dry land.

The new land animals moved into a world of plants. Large tropical forests flourished. The dead trees, instead of rapidly decaying, sank to the bottom of the oxygen-poor sub-tropical swamps, to be buried, compressed and cooked up first as peat, then coal, which we now burn to fuel our industrial growth. By the end of the Carboniferous period, roughly 300 million years ago, all the major families of plants had emerged, except the flowering plants (angiosperms) and cycad palms.

## THE END OF ANCIENT LIFE

At the end of the Permian period, 250 million years ago, there was one gigantic landmass on Earth – the supercontinent of Pangea – virtually stretching from the south to the north pole. The rest of the surface of the Earth was a single vast ocean. At this time, an unusually hot plume of hot mantle rock started to rise from deep within the Earth. The occasional appearance of these plumes is part of a cycle of deep convection, and is a natural consequence of the behaviour of the Earth's mantle (see Chapter 4). As the plume head reached the base of the overlying continent, it started to melt. In a period of less than a million years, an enormous volume of basaltic lava erupted on the Earth's surface. This is now preserved in Siberia near Noril'sk, and forms a volcanic pile over 3 kilometres thick in an area of about 2.5 million square kilometres. During the eruptions, huge quantities of dust, carbon dioxide, water vapour and sulphur dioxide were released into the atmosphere. The dust was trapped in the upper atmosphere, blocking out much of the sunlight, and the Earth darkened and cooled. The sulphur dioxide reacted with water to form sulphuric acid, which eventually precipitated as acid rain. The rain could have turned the surface of the Earth into an acid bath. This may have provoked the catastrophe in the oceans which occurred precisely at this time, when about 90 per cent of marine species went extinct. Corals, crinoids, brachiopods and ammonoids suffered heavy losses. The extinction also marked the final demise of the trilobites, which feature so prominently as fossils in older marine rocks. Life on land was also affected. It is estimated that more than 90 per cent of tetrapod (four-legged) land animal species, along with many orders of insects, went extinct. This episode – the end-Permian mass extinction – was the greatest single catastrophe in the history of life on Earth.

Some geologists believe that at least some of the end-Permian extinctions were due to longer-term changes in the global environment, including a massive drop in sea level which would have drained all the continental shelves, with catastrophic consequences for marine life. Certainly, the sea was nearly at its lowest recorded level, though the reason for this is unclear. Also, the clustering of the continents into a single land mass would almost certainly have had climatic consequences, possibly restricting ocean current systems so that the oceans became locally stagnant. It may be that longer-term environmental change set the scene for mass extinction, while the eruption of the Siberian lavas, and other volcanic eruptions which have left their mark in China, may have delivered the final *coup de grâce*.

## LIFE IN MIDDLE AGE

The end-Permian mass extinction marks a major turning point in both the evolution of life and the planet. Forces within the Earth were pushing at the edge of Pangea, as the ocean floor was dragged down into the Earth's interior. Huge coal deposits, which had accumulated in the preceding tens of millions of years all round the margins of the supercontinent, were thrust up and exposed to the atmosphere. As these deposits weathered, the vast quantities of the greenhouse gas carbon dioxide, locked up in the coal, were released to the atmosphere. In consequence, the climate heated up. Sea level also started to rise. Life took advantage of the warmer conditions. A new animal began to dominate the planet – the dinosaurs.

The dinosaurs comprise two orders of reptiles with a peculiar bone structure. In particular, the skull contains two large frontal holes and the joints and spine have a characteristic arrangement. The closest living relatives of the dinosaurs are birds and crocodiles. The diversification of dinosaurs really began to take off in the Jurassic, about 200 million years ago. An enormous range of dinosaurs evolved, from giants tens of metres long to small creatures a few tens of centimetres high. Some were docile grazing herbivores, and others were agile and aggressive carnivores. Most lived exclusively on land. There is evidence that they were capable of complex social behaviour. The fossilized remains of clutches of eggs, carefully arranged in nests, suggests that many dinosaurs reared their young. The pattern of fossil footprints, preserved in sandstones that were laid down by rivers, shows that dinosaurs moved in organized herds, walking on parallel tracks. A famous fossil, found in a Late Jurassic limestone quarry near Solnhofen in Bavaria, provides evidence that dinosaur-like creatures were not confined to the ground. This fossil, called *Archaeopteryx*, had feathery wings, and is considered to be the earliest bird, indicating that modern birds are the direct descendants of the dinosaurs. Interestingly, as birds are warm-blooded, this may provide support for the idea that some dinosaurs were warm-blooded. The other warm-blooded creatures in the landscape –

mammals – were mostly the size of shrews and rats, scuttling around and burrowing underground. Many were probably prey for carnivorous dinosaurs with their razor-sharp teeth and slashing claws.

For those who believe that the dinosaurs were mainly cold-blooded, the warm climatic conditions during the Jurassic and subsequent Cretaceous provide a reason why the dinosaurs managed to remain the dominant form of life on land. Unlike warm-blooded creatures such as the mammals, cold-blooded dinosaurs would not need to devote a large fraction of their energy input from food to maintaining their body temperature, and so could spend less time feeding. The large size of many dinosaurs would also have been a positive advantage, as this resulted in a relative decrease in the body surface area compared to body volume, making heat loss a less significant problem. In contrast, the small

*About 100 million years ago, in the Cretaceous period, the Earth was much warmer than today and large dinosaurs roamed the land. These dinosaur footprints are preserved in Cretaceous sandstones in Bolivia. Each footprint is about 20 centimetres across.*

*The tuatara, a reptile from New Zealand, is truly a living fossil. This one is about half a metre long. Nearly identical reptiles lived about 225 million years ago on the supercontinent of Pangea.*

warm-blooded mammals would have a major heat loss problem, because their body surface area is proportionately larger, compared to body volume, and maintaining a constant blood temperature requires a large energy input – hence the evolution of fur. Small mammals need to be hyperactive feeders. Thus, in the Cretaceous, the warm-bloodedness of mammals would have been an advantage only in marginal environments. This may explain the evolutionary success of the dinosaurs. But the price the dinosaurs had to pay for this success was that they were at the mercy of the planet's climate, which the geological record has shown is 'subject to change without notice'.

The existence of the supercontinent of Pangea for about 100 million years, overlying a large part of the Earth's mantle, suppressed the escape of heat from the mantle to the surface, acting as a sort of gigantic blanket. The mantle gradually heated up, expanding and doming upwards. This may have eventually forced the supercontinent apart, triggering the splitting of the northern part of Pangea (North America, Europe and Asia) from the remaining southern part (Gondwanaland) about 170 million years ago. Subsequently, Gondwanaland began to split up. In the early Cretaceous, 120 million years ago, Africa started splitting away from South America, and the South Atlantic began to form.

The rifting in Pangea created extensive new ocean floor. The new ocean basins, being young, were generally shallower than the older ocean basins (see Chapter 2). In addition, the ascent of a mantle plume beneath what is today the western Pacific, which caused intense volcanic activity between 120 and 80 million years ago, may also have pushed up the ocean floor. This way, large volumes of water were displaced on to the continents, creating extensive shallow seas. Here, planktonic organisms thrived. Some, with a calcium carbonate shell (Coccoliths), eventually accumulated on the sea floor to become chalk. Much of the organic matter was buried, gradually heating up to form the fabulously big oil fields found today on the continental shelves all over the world.

The split-up of Pangea, and subsequently of Gondwanaland, had a profound effect on the evolution of life, isolating whole populations of organisms.

Today, New Zealand preserves a so-called 'living fossil' – the tuatara – which is almost identical to small reptiles that lived in the Mesozoic on Gondwanaland. As the continents moved apart, they carried their cargo of floras and faunas into new climatic zones at different latitudes. Extinctions, including the loss of many dinosaurs, occurred as a result of this, but also new more adaptable species emerged, including flowering plants (angiosperms) 130 million years ago. In addition, the movement of the continents may have influenced the global climate. The middle Cretaceous was a particularly warm time, and there were forests close to the north pole. Perhaps new ocean current systems, established as sea ways were opened up between rifting continents, may have helped to carry equatorial warmth to polar regions.

## DEATH FROM ABOVE

Sixty-five million years ago, the long Cretaceous 'summer' came to an abrupt end. In some respects, there seems to have been almost a re-run of the events at the end of the Permian. But this time the Earth was dealt a double blow. A mantle plume welled up beneath India as it rifted away from the Seychelles Bank. Melting in the plume head resulted in a vast outpouring of lava in the Deccan region of western India over a period of less than a million years. As at the end of the Permian, volcanic eruptions on this scale must have had a very serious effect on the global climate. While this volcanic activity was going on, another catastrophe struck the planet.

The Earth does not orbit the Sun in isolation. Along with its planetary neighbours there are thousands of rocky fragments floating in space. Every so often one of these crashes into the Earth – many had early in the planet's history (see Chapter 8), and the scars of impacts are all over the Moon. And so it should be no surprise that a meteorite, about 10 kilometres in diameter, hit what is today Mexico 65 million years ago. Here, near Chicxulub in the Yucatán Peninsula,

the remains of the impact crater, 180 kilometres or more across, are preserved.

It has been estimated that the energy released during the Chicxulub meteorite impact was equivalent to the detonation of a megaton bomb on every square kilometre of the Earth's surface. In the crater, the Earth's crust, down to a depth of 30 kilometres, was shattered and locally melted. In fact, dating the time of melting gives the age of the meteorite impact. Dust particles were scattered all over the planet, blocking out much of the sunlight. The dust settled out on the sea floor as a thin layer of clay which can still be recognized today in sedimentary sequences from this period. The clay layer contains the tell-tale signs of the impact. It is enriched in the extremely rare element iridium, which is abundant in many meteorites. There are also microscopic droplets of once molten glass, and quartz with a particular fracture pattern characteristic of explosions or sudden impacts. The Chicxulub meteorite happened to hit an area which was underlain by gypsum deposits. These vaporized, releasing huge volumes of sulphur into the atmosphere. The effect of this on the atmosphere and oceans was probably much the same as large-scale volcanic eruptions, though compressed into a much shorter period. The sulphur reacted with water vapour, eventually precipitating as acid rain. It has been estimated that this could have made the surface waters in the oceans extremely acidic, with a pH as low as 3.

The fossil record clearly shows that life on Earth was devastated in a period of less than a million years, spanning the time of the Deccan volcanic eruptions in India and the Chicxulub meteorite impact. Many marine organisms, such as the ammonites and belomnites, went extinct. All land animals weighing more than about 25 kilograms were wiped out, including all the existing dinosaurs. This was a mass extinction second only to the great crisis in the evolution of life at the end of the Permian.

Judging by the effects of a much smaller meteorite which hit Siberia in 1908, the Chicxulub impact would have created fire storms in a region extending for thousands of kilometres, incinerating plants and animals. However, the worldwide extinction of many

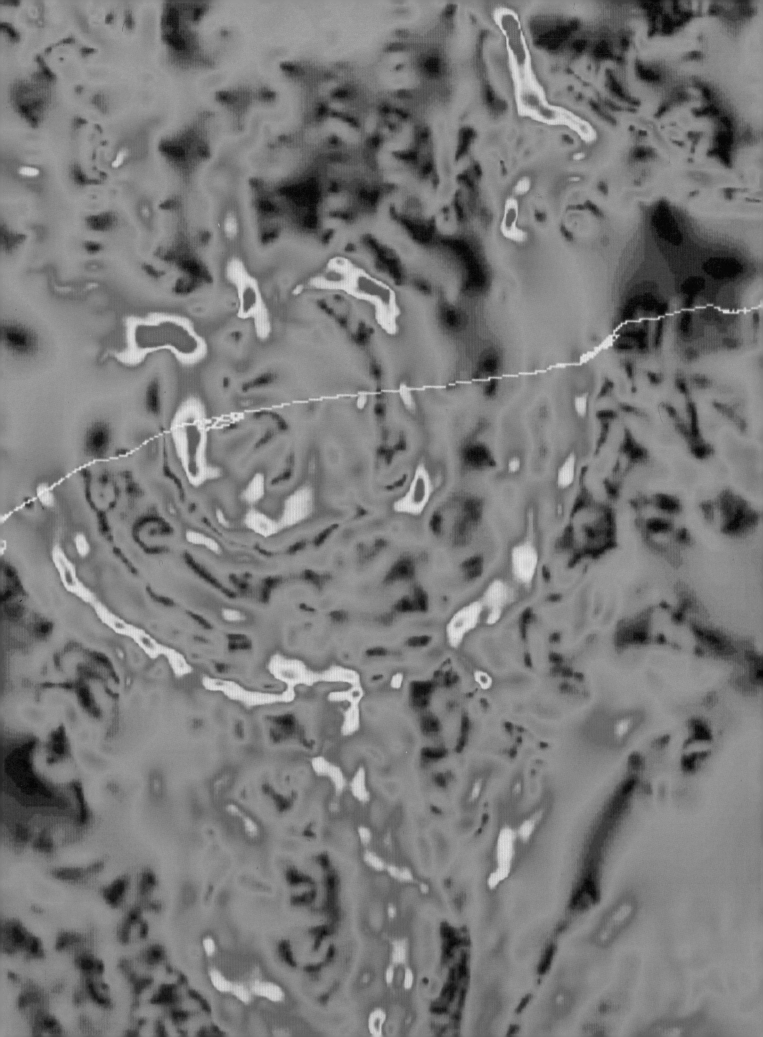

The earliest primates, from which we are descended, lived in the forests about 50 million years ago and must have looked a bit like this tiny tarsius monkey from Indonesia. With their forward-facing eyes and fingers, the early primates were agile clingers and leapers.

organisms may have required something more. The distribution of trees and other plants in the continents suggests that cooling of the global climate had started several million years earlier, and many dinosaurs and other land animals and plants had also become extinct before the end of the Cretaceous. It seems that both the Chicxulub meteorite and the Deccan eruptions were the final blow, tipping the balance between survival and extinction for many forms of life, which were already finding it difficult to cope with a longer-term cooling in the global climate. The final extinction of the dinosaurs was, from a human point of view, an extremely important event, because it cleared the way for the rise of the mammals and our own evolution.

*Left: Sixty-five million years ago, a large asteroid body about 10 kilometres across hit what is today the coast (white line) of the Yucatán Peninsula, near Chicxulub in Mexico. The remains of the impact crater, which is 180 kilometres or more wide, can be picked up in detailed surveys of the Earth's gravity in the region. The strength of gravity defines a series of concentric rings, colour-coded here in green, yellow and red.*

## THE THIRD AGE OF LIFE

It took about a million years for life to recover from the catastrophes which mark the end of the Cretaceous. The abundance of planktonic creatures, now preserved in deep sea sedimentary rocks, gradually increases in the rocks above the famous iridium-rich clay horizon. The global climate returned to warm and equable conditions, though much wetter than today. The tropical forests extended to middle latitudes and there were polar woodlands. Forests could become denser as they were no longer subject to the ravages of the dinosaur herbivores. Spread out in this world were the continental fragments of the once continuous Pangea and Gondwanaland, each with its cargo of mammals. These mammals were the descendants of a group of animals called cynodonts, which first evolved in the Triassic, about 230 million years ago. The essential features of living mammals are the ability to control body temperature (warm-bloodedness), production of milk to feed the young,

*About 20 million years ago, grasslands started to be a feature of the continents as the climate became drier at low latitudes. By 10 million years ago, the landscape in some continents may have looked much like this view of the Mara region of Kenya, teeming with grass-eating mammals.*

body hair, a complex ear, and a jaw structure which allows chewing. Most mammals go through their early stages of development inside the mother, fed via the placenta. In the Cretaceous, they were small generalists which had lived at the margins of the dinosaur world. Gradually, they started to move into the environmental niches vacated by the dinosaurs, and over a period of several million years evolved to larger sizes as they began to adapt to more specialized environments.

Ever since the break-up of Gondwanaland, India had been drifting northwards. By 55 million years ago, it had collided with the southern margin of Asia, and began to push up the great mountain ranges in the Himalayas and Tibet (see Chapter 5). Shallow water sediments, which had accumulated in the seas on the margins of India, were uplifted and exposed to the

atmosphere. These were rich in the remains of organisms and, as they weathered, the greenhouse gas carbon dioxide was released into the atmosphere. The global climate started to warm, resulting in more episodic rainfall so that the forest floor started to open up and smaller trees and bushes could flourish, providing food for many small ground-dwelling mammals.

In this new environment, mammals began to diversify: many of the present-day groupings (orders) of mammals evolved at this time, including bats, horses and early elephant-like mammals. Whales started to colonize the sea. Our own distant ancestors also appeared at this time. These were the early primates – rat-sized vertical clingers and leapers, similar to modern tarsiers, living in the tropical and subtropical forests. But their grasping hands with opposable thumbs, forward-facing sharp eyes and proportionately larger brains marked them out from other mammals. By the middle Eocene, 52 million years ago, global climate had reached its warmest point in the Cenozoic. At this time, the continued collision between India and Asia was causing large-scale mountain-building, and began to snarl up almost the entire global system of lithospheric plates, causing a reorganization of plate motions. This may also have triggered the onset of mountain-building and uplift in the Andes on the western margin of South America.

Mountain-building on a massive scale may have started to change the global climate in a new way. The shallow water sediments had weathered away, and now rocks from much deeper within the Earth's crust were uplifted and exposed. As these rocks weathered, in contrast to the weathering of shallow water sediments, carbon dioxide was sucked out of the atmosphere, reducing the concentration of this important greenhouse gas in the atmosphere. The inevitable result was global cooling, which has essentially continued until today, resulting in an overall drier climate.

By the Oligocene, 30 million years ago, the climate had become markedly more seasonal. The polar forests had gone and there was an ice sheet in Antarctica. The high-latitude regions became a tree-less tundra. The climate became even drier at low latitudes, and in the Miocene, about 20 million years ago, the savannah-like grasslands began to become a feature of the environment in the Americas. Grass has a long slender leaf, growing from the root upwards, and is exceptionally resistant to drought conditions. It can also regenerate itself very quickly, spreading out over the ground. Many mammals exploited this new form of food – specialized teeth and jaw structures evolved to enhance chewing. A complex digestive system also developed to deal with the roughage. Thus, once animals started relying on grass as a food source, they were locked into an evolutionary course which bound up their future survival with the survival of the grasslands. The scene was set for the latest chapter in the story of life – the evolution of the apes.

## THE RISE OF THE APES

In the language of taxonomy, human beings are primates, sharing the same order as apes, and the same family (Hominidae) as gorillas and chimpanzees. We are slightly more closely related to chimpanzees than gorillas, sharing over 98 per cent of our DNA with the former. Thus it is arguable that the difference between humans and chimpanzees is carried by less than 2 per cent of our DNA. In practical terms, the differences amount to our posture, body size, teeth layout, habitat and behaviour. The earliest 'humans' who shared these characteristics with us emerged about 5 million years ago in Africa, diverging from African apes. But the road to humanity started much further back in time.

About 26 million years ago, the first ape-like descendants of the early primates appear in the fossil record from northern Kenya. These are regarded as the earliest members of our own taxonomic family. The first well-studied example, called Proconsul, was also found in Kenya, dated to roughly 20 million years ago. Proconsul had hands with human proportions, but walked on four legs, living in the tropical rain forests. About 18 million years ago, the Afropithecines appear in the fossil record of East Africa. These were similar to living apes. The first ape-like fossils in Europe and Asia also date from this period — it seems that the land-bridge, created by the collision of the African continent with the European landmass, had allowed many of these to scramble out of Africa. The Afropithecines had teeth with a new feature in the evolution of primates: a marked thickening of the molar tooth enamel. The new dentition signals a change in diet to hard fruits and nuts, probably as a response to a less regular food supply in the progressively cooler and drier environment. The nuts are good for storage, but effective use of these stores requires a memory to find them again. There are even more profound evolutionary consequences. Mammals replace their teeth only once, and so they cannot survive much beyond the point when their

*Chimpanzees are our closest living relatives, sharing over 98 per cent of our genetic material. Hominids diverged from the ancestors of modern chimpanzees around 5 million years ago.*

teeth wear out. A strengthened enamel provides a means by which the average life-span can be prolonged. Hence, primates tend to live longer than other mammals.

There is a large gap in the fossil ape record between 18 and about 5 million years ago. In this period, the vast plateau of Tibet, occupying an area of several million square kilometres, had reached a height of several kilometres, pushed up by the northwards movement of India. It was a major feature on the surface of the Earth, and now had a significant impact on the local climate, deflecting major air currents and heating up the atmosphere at high levels. Some geologists believe that this caused a marked

strengthening of the Indian monsoon weather system, about 8 million years ago, which dominates the climate of so much of Asia today (see Chapter 5). The monsoon winds suck in moist air from the Indian Ocean during the summer months, focusing a large amount of rainfall into a relatively small part of the Earth during a period of a few months. The net effect of this is to make other parts of the planet drier.

Thus the strengthening of the Indian monsoon spurred on the longer-term shrinking of the tropical rain forests in Africa and the expansion of the grasslands. Almost unlimited sources of meat roamed these hot plains – herds of grazing antelope-like species, as well as predatory carnivores such as cats. An opportunity opened up for apes that could exploit the expanding grasslands. About 5 million years ago, the tree-dwelling apes took advantage of this opportunity, diverging from the chimpanzees. These are the first human-like apes or hominids, called the Australopithecines. They were relatively small-brained compared to humans, but walked on two legs. The famous 'Lucy' fossil (*Australopithecus afarensis*), found along with many other hominid fossils, in the Hadar region of Ethiopia, is the remains of a female about one metre high, with a hip structure similar to our own, who lived about 3 million years ago in a large family. Footprints of another Australopithecine are preserved in volcanic ash at Laetoli in Tanzania. This ash erupted from a nearby volcano about 3.7 million years ago, and formed a layer of mud after being soaked by a rain shower. Soon afterwards an elephant, a rhinoceros, some guinea fowl and a hominid walked across it, leaving a trail of footprints which were preserved as the ash hardened and was buried by subsequent volcanic fall-out.

The emergence of two-legged primates was a very important step on the evolutionary road to mankind. A change of posture from walking on four legs to two resulted in a knock-on effect, disproportionate to the anatomical change. Apes could run nimbly across the open savannah, reaching the next patch of forest without being mauled by a carnivorous cat, while the arms and hands were free for other tasks. Walking on two legs had other advantages as well. A four-legged ape, when in open country, exposes much of its body to the direct rays of the sun. This can be extremely harmful, causing the body to overheat if the animal exerts itself. An upright hominid exposes a much smaller cross-sectional area of himself to the sun. Thus organisms with a high metabolic activity, living in the intense heat of the open savannah, have a distinct advantage standing on two legs. Certainly, the ability to run on two legs and use the arms to wield a weapon are essential prerequisites of efficient hunting. Once hominids walked on two legs, they started to rely heavily on meat. This helped to fuel the metabolic activity necessary to support a large brain. But to be successful in hunting, they had to have a wide range of skills. Some scientists believe that this course of evolution led to a runaway development in behaviour and intelligence.

In general terms, the larger the brain, the greater the repertoire of behaviour it can exhibit. The key factor is the number and pattern of connections between nerve cells (neurons). The firing of neurons along a particular pathway of neuronal connections in the brain triggers a particular pattern of behaviour, be it the movement of a limb or a 'thought'. Some neurologists believe the brain behaves like a non-linear system – a very small change in the pattern of neuronal connections can lead to an extremely diverse repertoire of new behaviour in a large brain. This seems to be exactly what happened in the valleys and among the volcanoes of eastern Africa, created during the aborted rifting of the African plate along a north–south axis. The tectonic activity of the planet had produced an environment with rock shelters and rivers where hominids could thrive. Two million years ago, hominids living here, in the Olduvai Gorge of Tanzania, started to work pieces of volcanic rock. Evidence for tool-making over 2.5 million years ago has recently been discovered in Ethiopia. The Olduvai hominids, called *Homo habilis* or 'handy man', were clearly hunter–gatherers, killing and processing their food with pieces of volcanic obsidian. Their skulls show evidence for an enlargement of the region, called Broca's area, which is associated with speech in humans. Thus it is possible that *Homo habilis* could communicate to a limited degree through speech. A new species of hominid, called *Homo erectus*, is also

found at the Olduvai site, dated to about 1.7 million years ago. *Homo erectus* looked much more like us than other hominids, and was relatively tall, reaching 1.8 metres in height with a solid skull construction. It appears that about this time there was another migration from Africa, and *Homo erectus* fossils are scattered across Europe, Asia and China, mainly dating from the period between one million and half

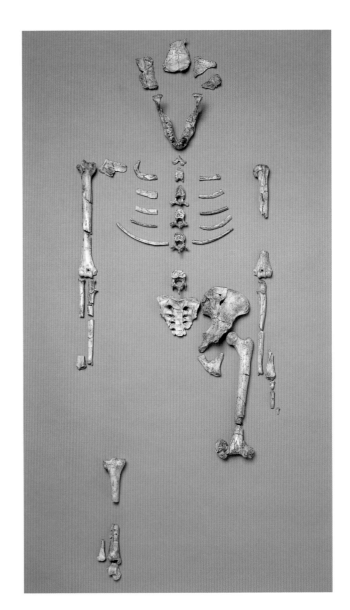

*The famous 'Lucy' fossil (Australopithecus afarensis), found in the Hadar region of Ethiopia, is the remains of a female about one metre high, with a hip structure similar to our own, who lived about 3 million years ago.*

*This is the oldest stone tool so far discovered, made by hominids in East Africa about 2.5 million years ago. Other animals use rocks as tools, but none fashion them for their own purposes.*

a million years ago. Some of the best examples were found near what is today Beijing, called Peking Man. Another important feature of *Homo erectus* is that he made elaborate tools, such as axes and knives.

During the last million years, as the Earth's climate oscillated violently between warm and cold periods, hominids needed a wider and wider range of behaviour to survive. By 200,000 years ago, Neanderthal man, named after the Neander Valley in Germany where the first fossils were found, had appeared on the scene, better adapted to coping with an Ice Age world. The Neanderthals were highly successful hunter–gatherers, with a complex social structure. They used fires and buried their dead. They could probably talk. But about 35,000 years ago, they dwindled in numbers and eventually died out. The reason for this decline is unclear. One suggestion is that the Neanderthals did not fish, and therefore were unable to benefit from this abundant and protein-rich source of food during the start of the last glaciation, when glaciers once again started to spread across northern Europe and the climate cooled dramatically. However, abundant fish bones are found at the shelters of a new species – *Homo sapiens*. The genetic evidence, based on the study of DNA, suggests that this species also came from Africa. Here, at warm equatorial latitudes, away from the worst excesses of the Pleistocene Ice Ages, *Homo sapiens* emerged before migrating out of Africa about 100,000 years ago. Subsequently, they replaced the Neanderthals and became the only members of the genus *Homo* on the planet: mankind.

*Left: This extraordinary trail of footprints, preserved in volcanic ash and found at Laetoli, Tanzania, records a journey on foot about 3.7 million years ago of two bipedal adult hominids (Australopithecus afarensis). The small round footprints on the right are those of an extinct three-toed horse.*

# A WORLD
# APART

• • • • • • • • • • • • • • • • • • • • • • • • • • • • • • •

Is the Earth unique, and if so, why?
To find an answer, scientists have had to
explore the Solar System, searching for clues
about our planet's birth. Uniquely amongst the
terrestrial planets, the Earth has retained liquid
water on its surface for over 4 billion years,
despite a steady increase in the Sun's heat
output. This water has had a profound
influence on the planet's geological activity,
as well as being a breeding ground for life. But
living organisms may have played a crucial role
in ensuring that liquid water exists on Earth,
linking the planet's geology and biology
tightly together.

*Looking into deep space – the region around the Horsehead Nebula in the*
*constellation of Orion, photographed with the 1.2-metre UK Schmidt telescope.*

In the previous chapters we have told a remarkable story of scientific revelation. This has been nothing less than the discovery of the workings of our planet. It is as though a watchmaker has looked inside an exquisite pocket watch. Opening the back reveals a mechanism of immense complexity – precision-made wheels within wheels, all interconnected and turning, powered by the unwinding of a spring. But the outward manifestations of this marvellous machine are simplicity itself: the hands turn in a regular fashion, passing at uniform intervals the minute and hour marks. And so the complex machinery of our planet, with all its connections between the deep interior and the surface, the overturn of tectonic plates and the coupling between the land, oceans and the atmosphere, also have a simple conclusion – the habitable world, so familiar to us, which has lasted over the aeons of geological time.

A natural question which anybody might ask, confronted with this scientific revelation, is: is the Earth unique and, if so, why? If we are to have a chance of answering this question, we need to have some idea of the essential character of the Earth. A biographer will naturally turn to the parents and early childhood of his subject. And so it is with geologists, who have tried to reach back to the origins of not just the Earth, but the whole Solar System, and even the Universe itself. Achieving such an understanding has been the outcome of a journey into the inner space of the world of atoms which are the fundamental components of our world. Physicists have gained sufficient experience of the building blocks of matter that, like geologists who can read the rocks, they can see a story in the atoms of the Universe.

It is only when we can view the Earth in the context of this much bigger picture that we can begin to question if and why the Earth embarked on its own unique course. Armed with the essential character of our planet, we can search space for signs of this in other worlds. In the last thirty years, space scientists have invested an enormous amount of effort in reaching the closest parts of our own Solar System. They have flown craft close to the Moon, Mars, Venus, Mercury, Saturn and Jupiter, and many of their moons, imaging the surfaces in great detail. Missions have actually landed on the Moon, Mars and Venus. None of these has turned out to be quite like the Earth, but there is evidence that they were similar in the past.

The contents of this chapter is really a joint enterprise between astronomers, cosmologists, nuclear physicists, space scientists and geologists. To begin with, we search for the common origin of the planets, from the creation of the Universe in the Big Bang to the early history of the Solar System. We then examine why the Earth developed in its own way, sustaining life.

## AN EXPANDING UNIVERSE OF ELEMENTS

The invention of the optical telescope in the seventeenth century allowed scientists to look in detail outside their world. Our planet is but one body among many which orbit a star, our Sun. The Sun and its orbiting planets, comets and asteroids comprise the Solar System. The Solar System is a small part of a galaxy of stars, and there are many galaxies, with their own stars and Solar Systems, spread out in the Universe.

Over 300 years ago, Isaac Newton discovered a glue which binds the Universe together and regulates its motions: the gravitational attraction of matter. Another of his important discoveries has helped to study this matter. Using a glass prism, he split sunlight into a multicoloured spectrum. The same experiment can be carried out for the light emitted by the highly compressed gases in the cores of other stars. Physicists now think of light in terms of waves with characteristic wavelengths – each colour corresponds to a range of particular wavelengths. White light contains all wavelengths. When white light passes through cooler gases which surround the hot core of a star, or form separate gaseous nebulae, or lie in the atmospheres of other planets, characteristic wavelengths of light are selectively removed, absorbed by the elements in the gases. When the remaining light reaches the Earth and is split, the missing wavelengths appear as black lines in what is called an absorption spectrum. Astronomers have catalogued the absorption spectra for the different elements and, by comparing these with those for light from bright objects in space, they have built up a picture of the range of elements in the Universe. There are 92 altogether, but over 99 per cent of all the visible matter in the Universe consists of the two lightest elements, hydrogen and helium.

So where did all the elements come from? The answer lies in another discovery made while astronomers were studying absorption spectra. The light rays coming from distant glowing nebulae have very similar absorption spectra patterns, but with one difference. The pattern remains the same, but the absorption bands for distant light sources are shifted towards longer wavelengths compared to the same absorption bands for nearby light sources. This is called the red-shift, first discovered by astronomers in the early part of this century. The only plausible explanation for the red-shift is that it is a 'Doppler effect', expected for light sources which are moving rapidly away from the Earth. The magnitudes of the red-shift require the distant nebulae to be accelerating away from the Earth at speeds of hundreds to thousands of kilometres per second. This suggests that the Universe is expanding. If it has always been expanding in the same manner, then somewhere between 7 and 20 billion years ago the Universe started as an infinitely small and dense point of matter, which exploded in the so-called Big Bang, in which space, time and matter were all created.

## THE CREATION OF MATTER

Cosmologists postulate that the creation of the elements was triggered by the energy released during the Big Bang, in a process called nucleosynthesis. They have some idea of what happened because physicists have built their own machines to create and destroy matter, unwittingly realizing the dreams of the alchemists of a bygone era. Experiments with these

machines have shown that the atoms of each element can be visualized as a heavy nucleus, made of subatomic particles called neutrons and protons, surrounded by orbiting electrons (see Chapter 1) – a sort of planetary system in miniature. In nuclear reactors and atomic accelerators, atoms are bombarded by fast moving electrons, protons and neutrons. Under these conditions, atoms can be split, releasing atomic energy. At extremely high temperatures, subatomic particles may fuse together to build heavier elements, again yielding huge quantities of energy.

The Big Bang theory postulates that the Universe at its inception was concentrated into a very small region. In these early conditions of extreme compression, the complex structure of an atom could not exist and matter consisted of naked neutrons. During the first moments of the Big Bang, this pressure was relieved and the neutrons spontaneously split into electrons and protons. Physicists have determined that this process will create roughly equal numbers of protons and neutrons in a few minutes after the Big Bang. And so the nucleus of the lightest element hydrogen, consisting of a single proton (an atomic weight of one), came into existence. Subsequent chance collisions between protons and neutrons created the heavier helium atom, consisting of two protons and one or two neutrons.

The Big Bang created a Universe of hydrogen and helium. But forces were unleashed which would subsequently build the remaining elements as the Universe continued to expand, and hydrogen and helium were scattered in nebulae of gas. Gradually, through the mutual gravitational attraction of mass in eddies and swirls of the gaseous nebulae, regions began to coalesce to form stars. As these stars contracted, incredibly high temperatures were reached, and the nuclei of hydrogen and helium atoms collided and fused. This process is still going on today, repeated many times within the hot cores of stars called red giants. Step by step, the fusion reactions have built up heavier elements in the periodic table. Iron, with an atomic weight of 56, seems to be the heaviest element that can be made this way. However, a star eventually exhausts its atomic fuel and huge amounts of gravitational energy are released as it falls in on itself, in the same way that falling water releases energy which can be harnessed to generate electricity. In these conditions of intense neutron and proton bombardment, elements heavier than iron can be created. The shock-waves of this collapse eventually cause what is called a supernova explosion – spectacular photographs of these have now been taken from the space observatory of the Hubble telescope.

The Hubble telescope has enabled astronomers to witness events, in distant parts of the galaxy, which may be the same as those which led to the formation of our own Solar System (see p. 218). Nebulae of gas, dispersed during a supernova explosion, locally seem to start condensing again through mutual gravitational attraction, separating out and starting to rotate. Eventually, the gas cloud forms a highly compressed central hot ball, fringed by a rotating disc of gas. This is most likely to have been the starting conditions for our own solar system, where the central gaseous ball became our Sun. Gradually, as the disc of gas cooled, the elements reacted with each other to form more complex minerals, and lumps of solid matter came into existence.

Gradually, lumps of matter, orbiting close to the central Sun, collided to form numerous small solid bodies (planetesimals), then rocky proto-planets, building up all the time to create eventually the rocky (terrestrial) planets (see p. 218). Not all the rocky matter was swept up into planets; numerous smaller rocky bodies either formed satellites or moons to the planets, or orbited the Sun as asteroids. At some stage, a violent shock-wave, created by ignition of atomic fusion in the Sun, blew more volatile gases such as hydrogen, helium, carbon, nitrogen and water vapour into orbit further away from the Sun. Here, planets accreted as gas giants or icy balls, made up mainly of hydrogen, helium and water ice. In this way, a solar system like our own could have been created, consisting of numerous asteroids and satellites, and the four terrestrial planets of Mercury, Venus, Earth and Mars in the innermost orbit around the Sun, and the outer gaseous and icy planets of Jupiter, Saturn, Uranus, Neptune and Pluto.

## EARTH HISTORY BEGINS

Speculating on the early stages of the formation of the Universe is really an occupation for astronomers, cosmologists and nuclear physicists. But these events set the scene for the evolution of the Earth – a subject of central interest to geologists – from the moment matter accreted to form a solid planet. A key date is the age of the Earth: this is the year zero of our history. But to find out this date, geologists have had to look at objects from outside the Earth.

Rocky fragments in the solar system regularly fall to Earth. Over the years thousands of meteorites, ranging in diameter from a few millimetres to a few tens of centimetres, have been found throughout the world. Antarctica is a particularly good place to find them, because they are preserved in ice, revealed like pebbles in the sand as the wind blows away their covering layer of snow. Their origins are variable. Some are thought to be fragments of the Moon or Mars, which were splintered off when giant asteroids hit them. Some are thought to be fragments which have been floating in space, orbiting the Sun since the creation of the Solar System. The commonest, called chondritic meteorites, form 86 per cent of all observed falls. These contain characteristic spherical structures or chondrules, typically 0.5 to 1.5 millimetres across, which probably cooled from molten droplets in space – this structure is not observed in rocks on Earth. One type of chondritic meteorite, called a carbonaceous chondrite, contains, among other elements, both carbon and up to 15 per cent by weight water – important ingredients, as it turns out, of both our planet and living organisms. The composition of this meteorite, except for highly volatile gases such as hydrogen and helium, is very similar to that of the surface of the Sun, determined by absorption spectroscopy. In fact, this is probably the non-volatile composition of all of the Sun, because convection will ensure that the Sun is well mixed. Thus, carbonaceous chondrites are thought to be condensed samples of the original

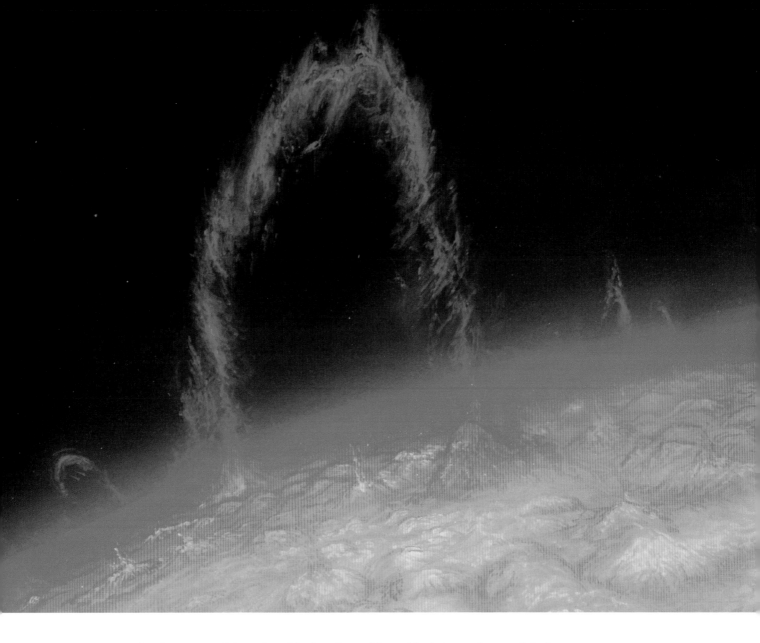

*Artist's impression of the surface of the Sun, where red flares of hot hydrogen plasma periodically erupt.*

nebula from which the Sun and the rocky planets, including the Earth, formed, containing all the essential components of these planets. This idea, called the chondritic Earth model, is the basis for many of our ideas about the early history of the Earth.

Radioactive elements in the chondritic meteorites, such as rubidium, sammarian and potassium, have been steadily decaying to their daughter elements ever since the matter in the chondritic meteorites first coalesced. Thus the abundance of the radioactive decay products can be used to date these meteorites; this shows that the chondritic meteorites have very similar ages, all being about 4.55 billion years old.

These are the oldest rocks so far discovered in the Solar System, and their ages are now considered to date from when solid matter crystallized in the solar nebula. In fact, geologists believe that from the moment this happened in the Solar System, fragments started to collide and stick together, building up all the time. This accretion continued until much of the matter had clumped together as planet-size balls. Thus the age of the chondritic meteorites is effectively the age of the Earth's birth.

The rock record on the surface of the Earth really begins about 3.8 billion years ago, when shallow seas lapped at the shores of a volcanic island, and iron-rich

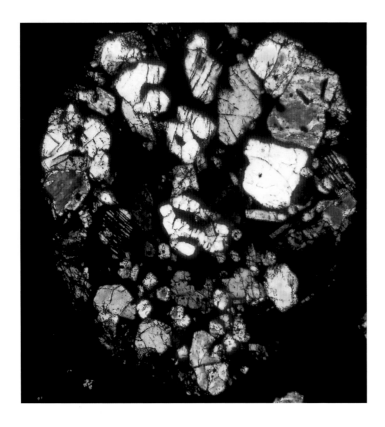

*This meteorite, called a carbonaceous chondrite, is a crystallized fragment of the solar nebula from which the Solar System formed. When viewed under the microscope (the field of view is about 1mm across) crystals of olivine and pyroxene are clearly visible (pink, yellow). The meteorite is also rich in carbon and water, essential components of life on Earth.*

sediments, now found in the Isua region of Greenland, were laid down (see Chapter 1). Nothing older has been found, except some individual minerals – about 4.0 billion years old – trapped in younger rocks. At first, it seemed as if a veil had been drawn over the first 500 million years or so of our planet's history. But this veil began to be lifted when geologists turned their attention to the Moon. The Earth and Moon are coupled together in space, held by a mutual gravitational attraction which forces the Moon to endlessly orbit the Earth. It has long been thought that the origin of the Moon is in some way bound up with our own planet. But no scientist, prior to the lunar missions, could have known that visiting the Moon would be an important step in unravelling the early history of the Earth.

## A GEOLOGIST ON THE MOON

In 1964 Harrison 'Jack' Schmitt joined the United States Geological Survey to work with Eugene Shoemaker on producing geological maps of the Moon. They used photographs of the Moon, taken with powerful telescopes. It was possible to work out a sequence of lunar events by studying the relationships between features on the lunar surface. For instance, if a small meteorite impact crater is found within a larger one, then the small crater must be younger. But, to date this sequence, they needed somebody to go and collect rock samples. In 1972, Schmitt landed on the Moon as part of the Apollo 17 mission. He is the only geologist to have set foot on another world. Schmitt spent three days on the Moon, going on a number of excursions in the lunar rover. Like a geologist back on Earth, he chipped off rock fragments with a geological hammer and also collected soil samples; 110 kilograms of lunar rock were brought back to Earth on this mission.

Compared to the Earth, the Moon is a dead place. It has a very weak magnetic field and can have only a small liquid iron core. This is also suggested by the Moon's average density, which is close to the density of the Earth's mantle (i.e. the Earth without its dense core). The lunar surface is pock-marked with craters, created when meteorites, ranging from a few centimetres to hundreds of kilometres across, slammed into it. Except for some possible small pockets of frozen water hidden in the depths of deep meteorite impact craters, the Moon is bone dry with virtually no atmosphere. The surface can be divided into rugged highlands (terrae) and smooth lowlands (maria). The highlands are composed mainly of a rock rich in the feldspar mineral, called anorthosite. This occurs on Earth where molten rock from the mantle has cooled slowly within the crust. The rocks in a lunar mare are basalt, which on Earth makes up much of the rock beneath the ocean floor, erupted from volcanoes.

The crucial discovery about the Moon was made when geologists started dating the lunar rocks, using

Harrison Schmitt is the only geologist to have visited another world. He landed on the Moon in 1972 on the Apollo 17 mission. During his 75-hour stay, Schmitt and his fellow astronaut Gene Cernan made three excursions in the Lunar Roving Vehicle and collected 110 kilograms of Moon rocks.

*The Moon is a dead world, pock-marked by impact craters. Lunar maria appear as dark smooth regions, surrounded by the bright highlands.*

a variety of radioactive decay systems as clocks. The lunar highlands have proved to be the oldest parts of the Moon. Eighty-five samples collected from this region have ages between 4.5 and 3.8 billion years. Most of these are older than any rocks known on the Earth. The oldest rock dates almost from the time of the accretion of the Earth. Schmitt describes how he skidded to a halt when he first noticed it because it had an unusual greenish colour. The fifty-two samples

collected from the lunar maria are fairly evenly distributed in age. These are lavas, erupted between 3.8 and 3.0 billion years ago.

Having established ages for the lunar rocks, geologists soon worked out the early geological history of the Moon. Initially, the surface of the Moon was molten, forming a magma ocean. Feldspar, which had crystallized from the magma ocean, floated to the surface and created a layer enriched in this mineral, which eventually cooled as anorthosite. The surface was also subjected to bombardment by huge meteorites. In the final phases of the most intense bombardment, about 3.8 billion years ago, craters over 1000 kilometres across were created. Between 3.8 and 3.0 billion years ago, the maria basalts erupted, filling meteorite craters with vast lava flows. These eruptions were probably triggered by further meteorite impacts while the Moon's interior was still hot enough to melt. Schmitt found the first tangible evidence of these eruptions – an orange–coloured ring of altered volcanic glass at the edge of one of the craters. It seems that when the meteorite struck, a fire fountain of molten rock spurted out. Around 3.0 billion years ago, from a geological point of view, the Moon essentially died. Thus when the rock record on Earth starts to become prolific, the record on the Moon ends. It had now lost so much heat that it was too cold to melt in subsequent impacts. Instead, its solid surface was pulverized to dust by meteorite impacts which have left their mark all over the lunar surface.

If we look at this early history of the Moon, then we see an image of what the early history of our own

*Lunar horizon showing the crater Copernicus in the distance, and two smaller prominent craters in the foreground. This picture was taken during the Apollo 12 mission.*

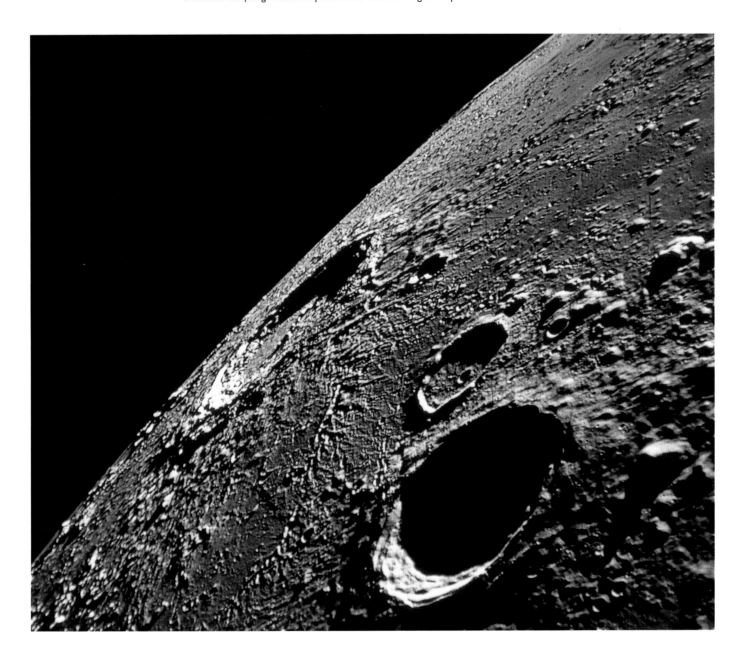

planet might have been like: an original all-enveloping magma ocean and a subsequent history of relentless meteorite bombardment, stirring up the surface of the Earth. But the Moon may have been much more than a silent witness to this early bombardment. Some scientists now believe that the Moon played an active part in the birth of the Earth we know today – this birth was nothing less than the creation of the planet's onion-like structure with a core, mantle, crust and atmosphere. The evidence for this comes from the coincidence between the age of the oldest rocks on the Moon and the timing of the formation of the Earth's internal layers.

## A TIMETABLE OF EARLY EVENTS

The simplest model for the origin of the Earth's layers is that they separated out from a homogeneous ball of matter. They form a series of shells arranged in order of progressively decreasing density from the centre outwards (see Chapter 4). Thus the high density central core, composed mainly of iron and nickel, is enclosed by a less dense mantle. In turn, the mantle is overlain by the even lighter crust, and the atmosphere forms the least dense outermost shell. In effect, the planet has evolved towards the most stable arrangement of its major internal divisions under the influence of gravity, in the same way that light oil tends to float on water, or less dense hot air rises.

Geologists are beginning to pin down when the Earth separated out into its distinctive internal layers. The story emerging is that these layers, except for the crust, formed early in the Earth's history, but not right at the very beginning. The biggest division is between the core and mantle – the core occupies a tenth of the Earth's volume, but is a third of its mass. A simple observation tells us that the formation of the Earth's core certainly occurred more than 3.5 billion years ago. The Earth's magnetic field is generated by fluid motions in the liquid outer core. Studies of the magnetism in 3.5-billion-year-old volcanic rocks,

found in South Africa and western Australia, strongly suggest that they were magnetized when they cooled from a molten state in the Earth's magnetic field.

Geologists can be even more precise about the age of the core. The calculation is based on an ingenious study of a number of unusual isotopes. These formed by decay of radioactive isotopes which are no longer found in the Solar System. The reason for this is that the parent isotopes decayed rapidly and had, geologically speaking, very short half-lives (see Chapter 1) – the isotopes of interest will decay to half their initial amount in less than 20 million years – and have now virtually decayed away. We know that this is the case, because the parent isotopes can be made and studied in man-made nuclear reactors. The time taken to become effectively extinct is, in fact, no more than a few half-lives, or not more than 200 million years. Here, we only attempt to give a flavour of the study of these isotopes. The important idea is that the segregation of the Earth into a core and mantle will redistribute the elements in the Earth: some will go into the core, some into the mantle. It is assumed that the rapidly decaying isotopes we are considering existed at the time of the creation of the Solar System, forged in the supernova precursor. Thus if the formation of the core took place before they had virtually decayed away, they would have been redistributed along with the other elements. In this case, one should see some sign of this in the abundance of their daughter products, which are stable and are found today. Two decay systems have proved particularly useful in this respect: the decay of an isotope of hafnium to an isotope of tungsten, and an isotope of iodine to an isotope of xenon. They show that the core had separated out later on in the accretion of the Earth, roughly 50 million years after solid matter first condensed in the Solar System – i.e. by about 4.5 billion years ago.

Surprising as it may seem, it is also possible to use the method described above to date when the Earth's present atmosphere came into existence. The composition of the present atmosphere is quite unlike that of nebula gas from which the Solar System formed. Its composition, except for the presence of oxygen, is in fact similar to that of gases emitted by volcanoes or hot springs. This strongly suggests that the present atmosphere is the accumulation of gases from the Earth's interior, released from molten rock during volcanic eruptions. The abundance of xenon-129 in the atmosphere and mantle suggests that most of the atmosphere had accumulated within 200 million years of the planet's birth – i.e. by 4.35 billion years ago. However, one important gas, oxygen, which today makes up about a fifth of the atmosphere, was not initially present. Oxygen reacts too readily with other elements in the Earth to naturally form a free gas. It was living organisms that found a way of generating oxygen gas, but it was not until about 2 billion years ago that oxygen started to accumulate in significant quantities in the atmosphere (see Chapter 7).

## SETTING THE WHEELS IN MOTION

The timetable of early events in the Earth's history gleaned from the daughter products of radioactive elements has a rather interesting implication. This timetable requires a rather hurried metamorphosis of the planet from coalescing lumps of the condensed solar nebula into the layered structure that exists today. But, even so, there was a slight delay. Accretion of matter started 4.55 billion years ago, but the layering did not exist until 4.5 billion years ago – a delay of roughly 50 million years. It is hard to imagine how such a metamorphosis could take place, on the timescales required, without virtually a complete melt-down of the planet. What would cause such a catastrophe? The tantalizing suggestion, referred to earlier, is that it was the creation of the Moon.

The idea is that early in the history of the Earth large asteroid bodies with similar chemical compositions to the Earth orbited the Sun on a path close to the Earth's orbit. About 4.5 billion years ago, one of these, about the size of Mars, eventually crossed the Earth's orbit, hitting the Earth with an oblique blow (see p. 218). The energy released during this impact melted both the Earth and the impactor.

This and the shock waves helped to trigger a chemical differentiation of the Earth. Volatile gases, which now make up much of our atmosphere – water vapour, carbon dioxide, nitrogen, xenon, argon and helium – separated from the molten matter, while the heavy iron–nickel fraction accumulated in the Earth's core. The volatile gases were trapped by the Earth's gravitational field to form the beginnings of an atmosphere. Eventually, water vapour condensed as an early ocean. However, some of the molten matter, with a composition similar to the Earth's mantle but without a volatile gas component, coalesced as a smaller body orbiting the Earth. This way, the Earth–Moon system was born and the Earth was literally kick-started into life. The Earth's mantle may have continued to feel the effects of other collisions with asteroids, but about 4.35 billion years ago, the Earth's atmosphere was here to stay.

The new ideas about the origin of the Earth's internal layering, atmosphere and the Moon suggest that the history of the Earth can be divided into two phases. To begin with, in the early phase the Earth heated up. This happened first when matter came together to build the planet, and would have been repeated during the collision of giant asteroids, such as the one which may have created the Moon. Heat was also created when iron and nickel collapsed into the Earth's centre, producing a molten core. The heat generated during this early time can be thought of as a store of primordial energy inherited by the Earth. The heat received by the Earth from the Sun can be ignored, though it is crucial for the temperatures in the atmosphere, because all of this is returned back into space (see Chapter 6).

In broad terms, the second phase in the history of the Earth, after the formation of the core, mantle and atmosphere, has been one of convection in the mantle, as hotter parts flow up to nearer the surface where they cool before sinking back again. Geophysicists have calculated that this cooling exceeds any further heating of the Earth's interior, and so overall the planet has become colder. The decay of radioactive isotopes throughout the history of the Earth – uranium, thorium and potassium being the most important – has caused further heating, but this

has become progressively smaller as the concentration of these isotopes has steadily dwindled away. In addition, the slow crystallization of the solid inner core from the remaining liquid core has been a source of energy, probably powering the dynamo which generates the Earth's magnetic field.

The steady cooling of the Earth since it acquired its early primordial heat has important implications. The mantle would have been less sticky (i.e. had a lower viscosity) in the early stages of the Earth's history when temperatures were generally higher. This stickiness tends to slow down mantle flow, so the mantle would have convected more vigorously when it was hotter and less sticky. As we have seen in Chapter 4, convection is the fundamental driving force for plate tectonics and volcanic activity at the Earth's surface. So, if plate tectonics was operating at this time, we would expect faster plate motions, smaller average plate sizes, and more extensive volcanic activity. The hotter mantle would have melted in larger quantities, producing a thicker oceanic-like crust than the present uniform 7 kilometres. A thicker crust would have made the lithosphere less dense, and so the plates may have had difficulty sinking back into the mantle again at subduction zones. For all these reasons, plate tectonics would not have been quite the same at this time compared to what we see today.

We have, at last, traced the history of the planet from its very beginnings to when, despite some differences, it starts to look like the Earth we know today. Given all the difficulties of looking so far back in time, we have a surprisingly good picture of the early growing pains of our planet. But how does this help us with the question we asked at the beginning of this chapter about the uniqueness of the Earth? Nothing we have described so far necessarily singles the Earth out from its nearest neighbouring planets. They too probably started at the same time, with a similar composition, and may have been knocked into shape by chance asteroid impacts, though none of them has an orbiting satellite like our Moon. We are in the same position as that biographer who, after extensive research into the childhood of his subject, concludes that in many ways it was unexceptional. We

## The birth of the Earth in the Solar System

Our Solar System was probably created when nebula from a supernova explosion **(a)** started to collapse gravitationally as a gas cloud rotating about a central gaseous ball **(b–c)**. About 4.55 billion years ago, matter crystallized and collided, and rocky bodies began to coalesce, orbiting around the Sun **(d)**. The innermost rocky bodies became the terrestrial planets, consisting of Mercury, Venus, Earth and Mars, with belts of asteroids. Initially, the Earth may have been a homogeneous ball with a surface magma ocean **(e)**.

About 4.5 billion years ago, an oblique collision between the Earth and a Mars-sized rocky body triggered the creation of the Moon **(f)**. In the process, the Earth melted and completed its internal segregation into an iron-rich core surrounded by the mantle **(g)**. Vapour released from the Earth's interior may have formed the beginnings of the atmosphere – water vapour subsequently condensed to form the early oceans. The lunar highlands crystallized at this time. The Earth was subjected to subsequent meteorite impacts, but by 4.35 billion years ago the atmosphere and oceans were probably here to stay and much of the Earth's surface was a cratered and volcanically active landscape **(h)**.

By 3.8 billion years ago, at the end of the phase of intense meteorite bombardment, convection in the Earth's mantle was slowly cooling the Earth. But this convection has also been the driving force behind plate tectonics and the creation of the continents **(i)**. There is also evidence for life on Earth at this time. Between 3.8 and 3 billion years ago, lavas erupted on the Moon, filling the lunar maria – subsequently, the Moon became a dead place. But the Earth has remained both geologically and biologically active.

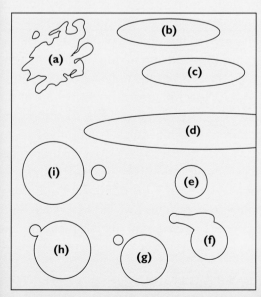

still have not found that elusive defining characteristic of the Earth. But perhaps it has been under our noses all the time? If we leaf through the previous chapters in the book, we see that one ingredient of the Earth seems to be underlying factor to so many of the features of the planet: the continents, plate tectonics, mountain ranges, the atmosphere, and life. This ingredient is water. We list below its influence in order of the chapters where this has been described.

- In Chapter 1 we describe the role of water in the formation of sedimentary rock. Water washes off the mountains carrying rock debris to the lowlands and sea. Here the debris accumulates in layers to form sedimentary rock.

- In Chapter 2 we describe the interaction between sea water and the ocean crust, which influences both the chemistry of the oceans and rocks. The circulation of cold sea water in the ocean crust helps to cool the lithospheric plate created at the mid-ocean ridge. Fabulously rich mineral deposits may form this way around the black smokers.

- In Chapter 3 we describe how water triggers the formation of the continents when it is carried down into the Earth's interior by the sinking ocean floor at subduction zones, acting as a sort of antifreeze in the mantle. In response, the mantle melts, yielding a water-rich magma which eventually cools and crystallizes as andesite or granite – granite is an important component of the continental crust.

- In Chapter 4 we describe how water has had a profound influence on plate tectonics. The presence of water weakens the rocks below and at the edges of the plates, allowing them to slide past each other easily. If this were not the case, the plates would soon jam up. As a result of plate tectonics, the surface of the Earth is constantly being rejuvenated and recycled.

- In Chapter 5 we describe how water penetrates rocks, weakening the minerals and drastically reducing the strength of the continental crust.

The weakened continental crust can be squeezed and pushed up to build large mountain ranges when continents collide. The surface of the continents is sculpted when liquid water runs off these mountains, on its way to the sea.

- In Chapter 6 we describe how water is an important component of the atmosphere. It stores solar heat, redistributing this over the surface of the Earth through the action of ocean currents. Water vapour, together with carbon dioxide, are 'greenhouse' gases. But the reaction between silicate rocks and water sucks carbon dioxide out of the atmosphere. So the presence of water has a dual effect on the temperature of the atmosphere.

- In Chapter 7 we describe the importance of water for living organisms. It is the stuff of life – it is more than 80 per cent of the weight of living organisms and is involved in the chemistry of life. It is probable that the interaction between surface water and volcanic activity in hot springs enabled life to start in the first place. Water provided the medium in which life first evolved.

The source of this water was the material in the solar nebula from which the Earth accreted. Large quantities of water would have been released during the early catastrophic degassing of the planet. In fact, the Earth was probably so rich in water that only a very small part would have had to melt to release the present volume of the oceans. Further collisions with icy bodies, such as comets, could even have added more water – an idea sometimes invoked for involved geochemical reasons. By the time the atmosphere stabilized, about 150 million years or so after the creation of the Moon, substantial oceans probably existed on Earth. Given the crucial role of water, the logical deduction is that plate tectonics (though possibly with much smaller plates than today), continents, and even life itself, came into being at this time. Direct evidence for the beginnings of all these is missing on Earth – possibly destroyed by the early intense meteorite bombardment recorded on the

*A dead planet: Mercury photographed by the Mariner 10 spacecraft in 1974. Impact craters are clearly visible.*

Moon. The oldest evidence in the rock record for life and oceans dates from about 3.8 billion years ago; for plate tectonics, about 3.5 billion years ago; and for continents, possibly 4.0 billion years ago, but certainly by 3.2 billion years ago.

There seems to be no shortage of water in the Solar System. Most of it is locked up in the outer icy planets. But except for the Earth, none of the rocky planets nearer the Sun has managed to keep liquid water on its surfaces. Scientists discovered this when they sent missions into space.

## VISITING OTHER PLANETS

Our nearest rocky planets, in order of increasing distance from the Sun, are Mercury, Venus and Mars. Mercury and Venus are closer to the Sun than the Earth, but Mars is further away. When viewed from Earth, the surface of Venus is obscured by its atmosphere, and Mars and Mercury are too far away to be seen clearly. Thus the results of the various

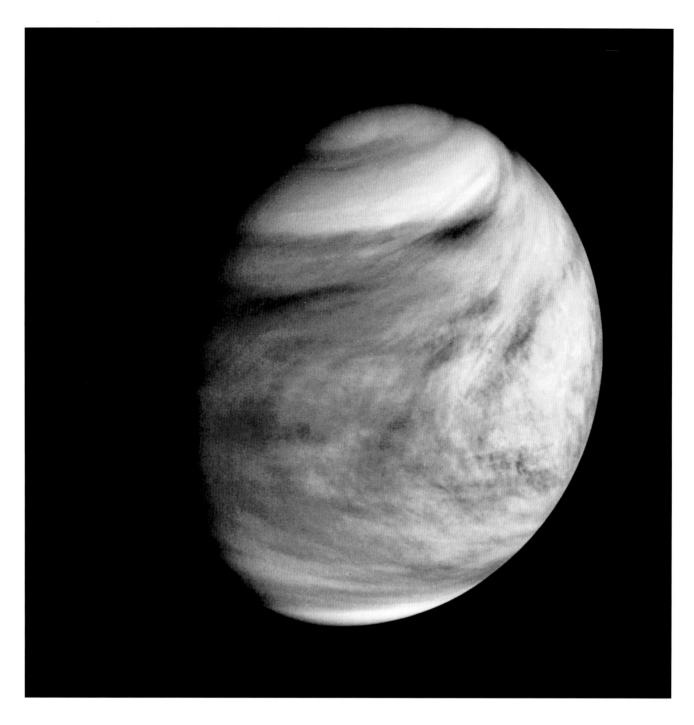

*Venus photographed by the Pioneer orbiter. Venus is the closest planet to the Earth and the second major planet from the Sun. The yellow-white cloud cover permanently obscures the surface.*

Russian and American unmanned missions to these planets have truly been a surprise.

Mercury is only slightly more than a third of the Earth's distance from the Sun. It is also the smallest of the rocky planets, with a diameter only about a third of the Earth's. No spacecraft have landed on Mercury, but in 1973 and 1974 the *Mariner 10* craft flew close by three times, revealing a lunar-like terrain. The meagre spectroscopic data suggests that the surface of Mercury is essentially basalt. Mercury has virtually no atmosphere and surface temperatures vary from 350°C in the day to −180°C at night. It seems to be a hostile, barren and cratered world, with a much higher average density than the other rocky planets. The high

*A perspective view of the surface of Venus showing the colossal Maat Mons volcano which rises about 8 kilometres above the surrounding regions. The image has been created from radar data collected by the Magellan spacecraft. The colours are based on the colours recorded by the Russian Venera 13 and 14 spacecraft.*

density of Mercury suggests that it has a large dense iron core, perhaps because a giant impact stripped away much of its less dense mantle.

In many respects, Venus is the Earth's twin. It is nearly the same size and has a very similar bulk density. However, it is only three-quarters of the Earth's average distance from the Sun. In the late 1970s, and subsequently in the early 1990s, the American spacecraft *Pioneer* and *Magellan* orbited Venus. *Magellan* measured the surface height of the planet with great precision and produced radar images. This made it possible to construct the first accurate

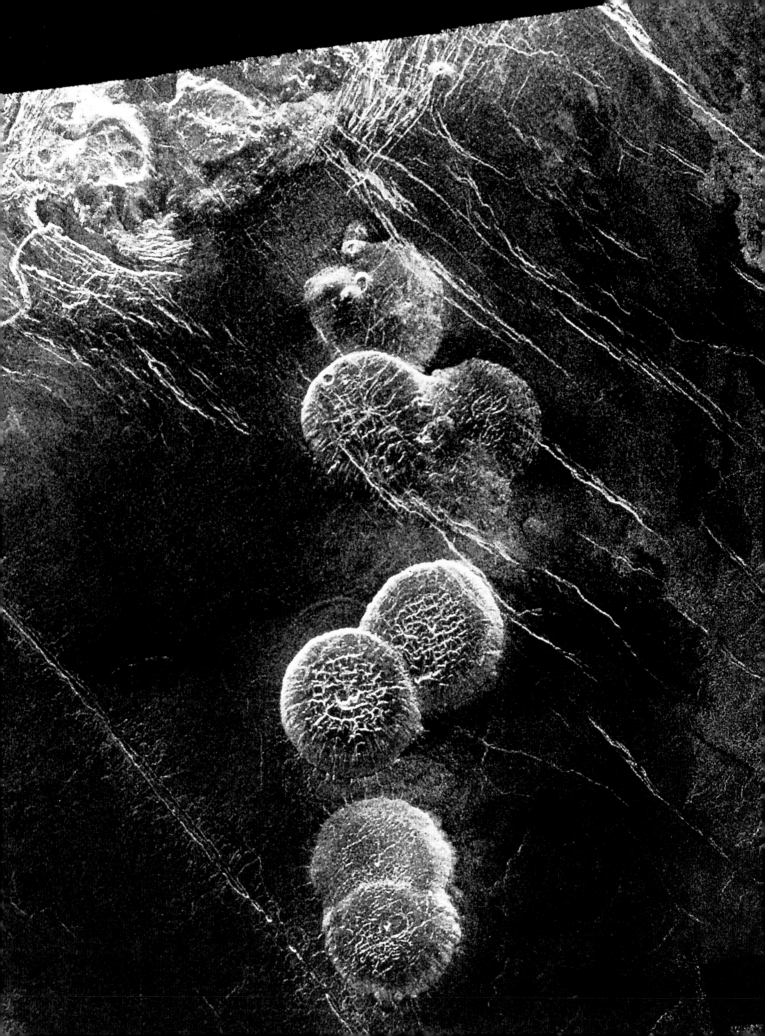

*These extraordinary photographs of the surface of Venus were taken by the Russian spacecraft* Venera 13 *in 1982 (parts of the spacecraft are visible). They show a rocky landscape strewn with fragments of volcanic rock. The surface temperature is over 400°C.*

topographic maps of almost all the surface of Venus. This looks quite unlike that of Earth, made up of huge volcanoes, lava flows and terrains with strange patterns. In places it is crinkled, as though squeezed together by forces inside the planet. Elsewhere, there are rifts and long linear mountains reminiscent of the mid-ocean ridges on Earth – together these suggest stretching and spreading of the surface. In 1982, the Russians managed to land the *Venera 13* and *14* craft on Venus. They transmitted back to Earth pictures of a

*Left: Radar image of the surface of Venus taken by the* Magellan *spacecraft in 1990. Several volcanic domes, each about 25 kilometres across and 750 metres high, were created by eruptions of sticky lava. Cracks, which formed when the lava cooled, are clearly visible.*

flat surface which seems to be made up of slabs of basaltic lava. They also found a surface hotter than molten lead, at an average temperature of 465°C, and an atmosphere made up almost entirely of carbon dioxide with a crushing pressure nearly 100 times that on Earth.

Venus may still be volcanically active. There are signs of lava flows which are similar to freshly erupted flows on Earth. Also, the 1990s *Pioneer* mission detected increased levels of sulphur dioxide in the Venusian atmosphere, compared to earlier 1970s measurements. The most likely explanation is that the sulphur dioxide was released during a recent volcanic eruption. However, meteorite impact craters are virtually everywhere. To account for the large number

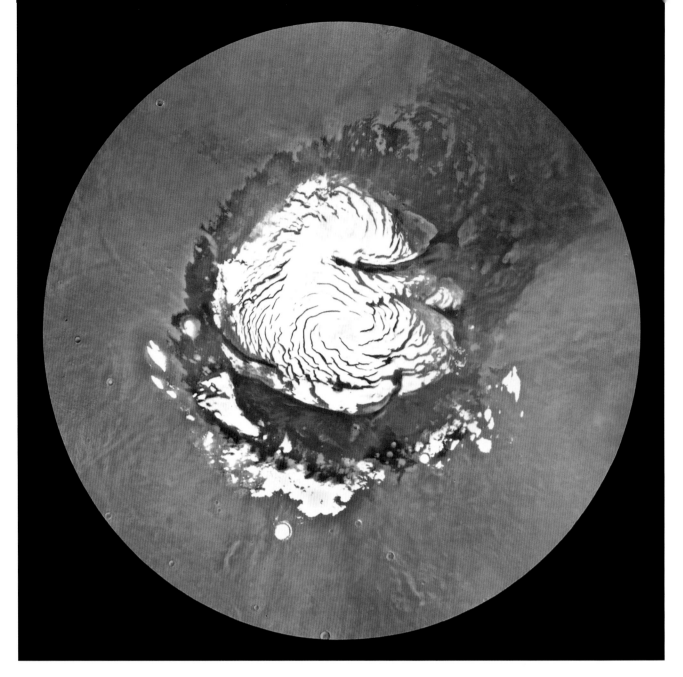

of these, given the likelihood of a collision, most of the present surface of Venus must have been in existence for at least 300 million years – ten times longer than the large area of ocean floor on Earth. Also, zones of both stretching and compression do not seem to link up in any simple way to outline large rigid plates which are so characteristic of the Earth's surface. Taken together, these observations seem to suggest that plate tectonics does not operate presently on Venus, at least not in the way we see it on Earth.

Mars is one and a half times the Earth's distance from the Sun, and has a diameter about half that of the Earth. In 1976, the two *Viking* spacecraft landed on nearly opposite sides of the planet, and most recently, in 1997, another American mission, *Mars Pathfinder*, equipped with a small robotic rover, touched down on Mars. Mars has an atmosphere 100 times more rarified than the Earth's, made up almost entirely of carbon dioxide. The average surface

*Left: The northern polar region of Mars, imaged by the* Viking *probes. A white ice cap, consisting of both frozen water and frozen carbon dioxide, occupies the centre of the scene.*

*Below: A* Pathfinder *photograph of the surface of Mars showing part of the craft and a boulder-strewn red landscape, not unlike that found today in many desert environments on Earth. However, the surface temperature of Mars is about −50°C and the atmospheric pressure is a small fraction of that on Earth.*

temperature is −50°C, cold enough to freeze carbon dioxide at the poles, creating the white polar caps. The images transmitted back to Earth reveal a surface that looks similar to parts of the Sahara Desert, with thousands of rocky fragments strewn over a sandy plain. Sand dunes are clearly visible, as well as patches of carbon dioxide frost. The true colours of these remarkable pictures show that the surface of Mars is indeed red, most likely stained by iron oxides. Raging dust storms which occasionally sweep across this surface explain why the appearance of Mars, seen from Earth, often seems to change.

Perhaps the most remarkable feature is the network of sinuous channels, very similar to river channels on Earth. Sometimes there are huge scoured plains, as though the landscape had been overrun by a flood – a great deal of water must have coursed across the surface of Mars early in its history. Any water left on Mars now lies frozen at the north pole or in the soil. Evidence for this is sometimes found in the ejecta of meteorite craters, where the energy released during impact has locally melted the water and liquefied the ground, creating characteristic fluid-like structures. An old water channel was chosen as the landing site for the recent *Pathfinder* mission. If the jumble of boulders found here have been transported large distances by the once flowing water, then they could be a good sample of the variety of Martian rocks. Indeed, one of the boulders examined by the

*Pathfinder* rover has already proved to be interesting. It is rich in silica, a characteristic of rocks on Earth which have undergone some sort of differentiation. There are signs of volcanoes on Mars, many times greater than those on Earth, and it is possible that some are still volcanically active. The colossal Olympus Mons volcano stands 26 kilometres above the general land surface, and is 500 kilometres across at its base. At the summit is a conspicuous depression (caldera) about 50 kilometres across.

## COCOONED IN SPACE

The exploration of the other terrestrial planets has clearly shown that none of them is geologically active in quite the same way as the Earth. None shows convincing signs of plate tectonics, or has the two clear types of crust – oceanic or continental crust – found on Earth (see p. 229). And none has liquid water, or an atmosphere like the Earth's, or living organisms. So we can say that the Earth is unique, at least in the Solar System, though there may be other solar systems with Earth-like planets. But why? It seems that though they may have started out in a similar way to the Earth, the other terrestrial planets ended up very differently. For example, there is no strong reason why plate tectonics may not have operated on these planets in the past. There was also once liquid water on Mars, and there is geochemical evidence that the early atmosphere of Venus was moist. So what happened? If we recall our ideas about the importance of water, then the fate of the other terrestrial planets may be tied up with their failure to preserve a suitable atmosphere capable of containing liquid water. As we shall see, both the size of the planet and its distance from the Sun seem to play a role in this.

Ever since the bulk of the Earth's atmosphere originally formed, during a catastrophic phase of degassing of the planetary interior, it has been constantly topped up with gases emitted during subsequent volcanic eruptions. This has been possible because the Earth's interior is hot enough to occasionally melt. Mars and Mercury, being so much smaller than the Earth, have cooled so much that volcanic activity is rare or non-existent, and they have no way of replenishing their atmospheres. One way that planets lose their atmospheres is through hydrogen fluxing, which causes a steady leakage into space. This happens when there is water vapour in the upper atmosphere which dissociates under the influence of solar ultraviolet radiation, creating 'free' hydrogen: the hydrogen, being light, acquires sufficient velocity from the ambient thermal energy to escape from the gravitational pull of the planet. In fact, the hydrogen reaches supersonic velocities, creating a wind which can literally blow other atmospheric gases away. But for hydrogen fluxing to be significant, there has to be both intense ultraviolet radiation and a sufficient amount of water vapour in the atmosphere. It is likely that Mercury, being so close to the Sun, lost its atmosphere very early in its history.

Venus also must have been subject to intense hydrogen fluxing. It receives enough solar radiation, being closer to the Sun than the Earth, to ensure that if there was once surface water on Venus, the atmosphere would have been warm and moist. Emissions during volcanic eruptions would have added carbon dioxide. These conditions contain the seeds of the destruction of the early Venusian atmosphere. The moist atmosphere would have had a strong greenhouse effect – water vapour is an effective greenhouse gas – pushing the surface of Venus into a vicious spiral of warming. As oceans evaporated in the warm conditions, more moisture was transferred to the atmosphere, which led to further warming, and so forth. But both water vapour and other gases were on a one-way journey to space, plucked out of the atmosphere by hydrogen fluxing. Eventually, the oceans had completely evaporated and the surface of Venus was dry. While there was surface water, atmospheric carbon dioxide could react with the silicate rocks on the surface of Venus in the so-called Urey reaction (see Chapter 6), and carbon dioxide was removed from the atmosphere and locked up in rocks or the oceans. But when the surface of Venus became dry, the Urey reaction could no longer take

MERCURY    VENUS    EARTH    MOON    MARS

*Mantle*

*Core*

*Mantle*

*Core?*

*Mantle*

*Outer core*

*Inner core*

— 2890km

— 6370km

0

*Mantle*

*Core?*

*Mantle*

*Core*

There are both differences and similarities between the Earth, the other terrestrial planets and the Moon. All their interiors consist of a central iron-rich core surrounded by the mantle. But the surfaces of Mercury and the Moon are cratered and dead places – their interiors are probably relatively cold as well. Mars is cratered, but shows signs of past volcanic activity – its interior is probably much cooler than the Earth's. The surface of Venus is hot, well over 400°C, and is volcanically active; Venus probably has a hot interior like the Earth's.

place. Rocks, with their store of carbon dioxide, when eventually buried and heated up by the geological activity on Venus, permanently lost their carbon dioxide to the atmosphere, and the level of carbon dioxide in the Venusian atmosphere rose dramatically. In fact, practically all the carbon dioxide on Venus now seems to be in the atmosphere. The 'greenhouse effect' of the high level of carbon dioxide (300,000 times the pressure of carbon dioxide on Earth) has kept the Venusian surface at its present high temperature of over 400°C.

The early history of the Martian atmosphere was probably similar to that on Venus and the Earth. There was clearly liquid surface water, and wet weathering of silicates could occur, locking up carbon dioxide in the rocks through the process of the Urey reaction. However, the smaller mass of Mars, compared to the Earth, resulted in more rapid cooling of its planetary interior. Thus, the intensity of surface volcanism declined early on and has been insufficient to make up for the carbon dioxide lost through the Urey reaction from the Martian atmosphere. Also, the low influx of solar radiation, because of Mars's distance from the Sun, suppressed hydrogen fluxing and kept the atmosphere cool and dry. Mars was pushed into a spiral, completely the reverse of that on Venus. The low concentration of greenhouse gases – water vapour and carbon dioxide – promoted cooling. Eventually, the temperature dropped below the freezing point of carbon dioxide, and all surface water and most of the remaining carbon dioxide on Mars were locked up as ice.

## A NATURAL THERMOSTAT

It seems that the atmospheres of the terrestrial planets, except the Earth, veered away from a moist and temperate environment towards far more extreme conditions, quite early in their histories. Both their

## THE EARTH'S ATMOSPHERE AND CARBON CYCLE

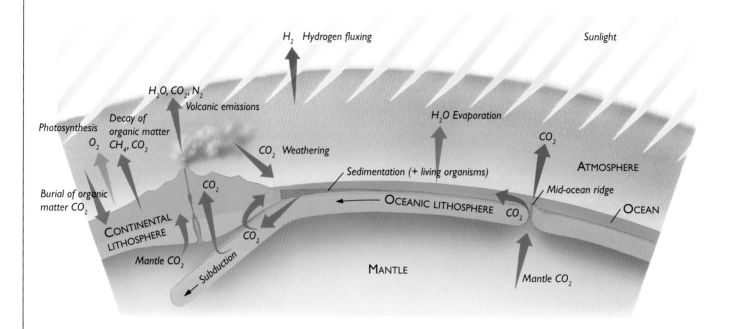

H₂ Hydrogen fluxing — $H_2$ Hydrogen fluxing

Sunlight

H₂O, CO₂, N₂ — $H_2O, CO_2, N_2$
Volcanic emissions

Photosynthesis
O₂ — $O_2$

Decay of organic matter
CH₄, CO₂ — $CH_4, CO_2$

$CO_2$ Weathering

$H_2O$ Evaporation

$CO_2$

ATMOSPHERE

Burial of organic matter $CO_2$

$CO_2$

Sedimentation (+ living organisms)

Mid-ocean ridge

OCEANIC LITHOSPHERE $CO_2$

OCEAN

CONTINENTAL LITHOSPHERE

$CO_2$

Mantle $CO_2$

Subduction

MANTLE

Mantle $CO_2$

Carbon is an important component of our world. It is in living or dead and buried organisms, in the atmosphere (as carbon dioxide), and in the crust (mainly as carbonates) and mantle. There is a flow of carbon between the various places on Earth where it is stored – this is called the carbon cycle. Both geological processes (plate tectonics, volcanic activity, sedimentation, weathering and erosion) and biological activity (growth or decay of organisms) control this cycle. In the process, the level of carbon dioxide in the atmosphere may rise or fall, changing the climate because carbon dioxide is an important greenhouse gas.

distance from the Sun and their intensity of volcanic activity must have played a role in determining this course. Once liquid water ceased to exist, the surfaces of these planets became far hotter or colder than the Earth. Venus, being so hot, almost certainly has the vigorous internal convection needed to drive plate tectonics but, because the surface is dry, it may lack the lubricating effect of water. This, and also the high surface temperature, may prevent plate tectonics operating in the way it does on Earth. Because of its size, the Martian planetary interior cooled more rapidly than the Earth's, and became too cold for the convection needed to drive tectonic plates. So, perhaps the unique history of the Earth is because it was just the right size and distance from the Sun.

But there is one other hurdle that the planets have had to contend with. An analysis of the energy yields from the atomic fusion of hydrogen in the Sun, which provides the solar power, suggests that the intensity of the Sun's radiation has increased by 20 to 30 per cent since the formation of the Solar System. Changes of this magnitude would be expected to have a catastrophic impact on the Earth's atmosphere, which clearly has not been the case. This is sometimes called the weak Sun paradox. Conditions on the Earth have stayed remarkably constant through the aeons of geological time. The presence of liquid water on Earth for most of its history – since 3.8 billion years ago, at the very least, there is a record of water-deposited sediments – tells us that the average temperature of the Earth's surface has remained between roughly zero and 100°C. Strictly speaking, this narrow temperature range depends on the Earth maintaining its present atmospheric pressure. Even so, the

*Mars photographed by the Viking orbiter. The centre of the scene shows the Valles Marineris canyon system, over 2000 kilometres long and up to 8 kilometres deep. The Tharsis volcanoes (dark spots), each about 25 kilometres high, are visible on the far right. Ancient river channels snake their way towards the top of the picture.*

*The blue planet – this view of Earth from space, centred on Africa, clearly shows the abundant water on the surface of our planet, either as a liquid or vapour, locked up in the deep blue oceans and the white swirling clouds.*

permissible temperature range is very small in comparison to the temperatures found on the surfaces of other planets, which range from $-230°C$ to well over $400°C$.

Greenhouse gases help to keep the Earth's atmosphere warm – greater concentrations enhance the warming effect. So, one resolution to the weak Sun paradox is that the level of these gases has progressively decreased throughout the history of the Earth, compensating for the increase in solar output and keeping the surface of the Earth within narrow temperature limits. The level of the important greenhouse gas carbon dioxide is controlled by a number of factors. Volcanism tends to add carbon dioxide to the atmosphere. Wet weathering of silicate rocks, via the Urey reaction, removes it. This reaction is promoted by both higher atmospheric temperatures and the activity of plants. Living organisms also extract carbon dioxide, putting it in their bodies or shells as carbon. The mass of carbon contained in living organisms today is over four times the atmospheric mass of carbon (mainly in the form of carbon dioxide); and the mass of carbon in rocks, there because of the activity of living organisms, is 100,000 times this. Thus an increase in the biomass of only 25 per cent, or a small change in biological activity, could strip the atmosphere entirely of carbon dioxide. Subduction, as part of plate tectonics, takes both rocks and the remains of living organisms down into the Earth's interior where they are heated up. In the process, the carbon they contain is released and eventually returned to the atmosphere as carbon dioxide. So we have in effect a complex system made up of temperature-sensitive chemical reactions and full of checks and balances, which seems to have subtracted or added carbon dioxide to the atmosphere in just the right way to maintain roughly constant surface temperatures.

The system outlined above begs an intriguing question. Could it operate if any part was missing? For example, if we took away the effects of living organisms, would it work? The answer to this may be no. Living organisms rely on chemical reactions which are very sensitive to temperature. This sensitivity may be the ultimate thermostat on Earth. What we have really described here is a partnership between geological and biological activity. The two act together and each is dependent on the other. This is a truly amazing conclusion. The Earth gave birth to life in the first place, but life has helped to sustain the geological life of the planet.

We have now gone a long way towards answering the question we asked at the beginning of this chapter: is the Earth unique and, if so, why? We singled out the presence of liquid water as the defining character, making it unique in the Solar System. The survival of liquid water on the planet's surface is determined by many things. The distance of the Earth from the Sun has proved important. The size and composition of the Earth itself have a role, enabling the Earth to cool in its own particular manner with profound influences on both the evolution of the atmosphere and life. Pure chance may also have played its part, so that by some miracle the Earth, at least in the last 4 billion years, has avoided being hit by a giant meteorite large enough to blow off the atmosphere and wipe out all life.

A recent suggestion by NASA and other scientists of life on Mars, at an early stage in Martian history, gives us a new perspective on the role of life on Earth. If there was life on Mars, it is unlikely that it survived long. The creation of life in the Universe may be a common occurrence. But its survival over the aeons of geological time may well be extraordinarily rare. However, it is unlikely that life on Earth will be able to pull off the trick of survival for much more than a further billion years. By then, the Sun will be giving off so much solar radiation that even the Earth's well-tested thermostat will fail to function.

That this probable demise is so far off in the future, or that the Earth has managed for so long to maintain suitable conditions for life, are not grounds for our complacency. We, as human beings, are fast becoming a powerful force on the planet. We are capable of changing some of its geological activity – the flow of rivers and the erosion of the land, to name but two processes which we can influence – as well as affecting the workings of the atmosphere and oceans. We must use our power wisely. This we can do only if we understand fully how the Earth works. The chapters in this book have described the beginning of our quest for understanding. Though we have made great progress – the basic framework of our knowledge about the Earth is unlikely to change – there are many details still to discover. It is from these details that we can learn how to live sustainably on the planet. That should be our goal in the future.

# Select Bibliography

Numerous books, articles and original research papers were consulted during the writing of this book. In a scientific publication, these would have been referred to in the main text at the appropriate places. However, in order to make the subject matter less daunting to a general audience, we have not included references, although we owe an enormous debt to the authors of these works. For those readers who wish to follow up on some of the topics dealt with in this book, a selected reading list is given below.

### General Reading

A. Cox (Ed.), *Plate Tectonics and Geomagnetic Reversals*, Freeman & Co., San Francisco, 1973

R. Osborne & D. Tarling, *The Viking Historical Atlas of the Earth*, Penguin Books, London, 1995

E. J. Tarbuck & F. K. Lutgens, *Earth Sciences* (8th ed.), Prentice-Hall, New Jersey, 1997

T. H. Van Andel, *New Views on an Old Planet*, Cambridge University Press, Cambridge, U.K., 1994

### Chapter 1

J. Burchfield, *Lord Kelvin and the Age of the Earth*, Macmillan Press, London, 1975

G. B. Dalrymple, *The Age of the Earth*, Stanford University Press, California, 1991

S. J. Gould, *Time's Arrow, Time's Cycle*, Penguin Books, London, 1991

### Chapter 2, 3, 4 & 5

W. Glen, *The Road to Jaramillo*, Stanford University Press, California, 1982

A. Hallam, *A Revolution in the Earth Sciences*, Clarendon Press, Oxford, 1973

H. W. Menard, *Islands*, Scientific American Books, New York, 1986

H. W. Menard, *Ocean of Truth – A Personal History of Global Tectonics*, Princeton University Press, Princeton, 1995

G. Ranalli, *Rheology of the Earth* (2nd ed.), Chapman & Hall, London, 1995

J. Verne, *Journey to the Centre of the Earth*, Penguin Books, London, 1965

### Chapter 6

B. G. Anderson & H. W. Borns, *The Ice Age World*, Scandinavian University Press, Oslo, 1994

T. E. Graedel & P. J. Crutzen, *Atmosphere, Climate, and Change*, Scientific American Books, New York, 1995

J. Houghton, *Global Warming*, Lion Publishing, Oxford, 1994

### Chapter 7

S. J. Gould (Ed.), *The Book of Life*, Ebury Hutchinson, London, 1993

J. W. Schopf, *Major Events in the History of Life*, Jones & Bartlett, Boston, 1992

### Chapter 8

W. S. Broeker, *Building a Habitable Planet*, Lamont-Doherty Geological Observatory of Columbia University, Eldigio Press, New York, 1985

N. Henbest, *The Planets – Portraits of New Worlds*, Penguin Books, London, 1994

M. Ozima, *Geohistory – Global Evolution of the Earth*, Springer-Verlag, Berlin, 1987

# References

### Chapter 1

page 14 'On us who saw...' John Playfair, *Transactions of the Royal Society*, 4, 39, Edinburgh, 1803

page 16 'no vestige of a beginning...' James Hutton, *Transactions of the Royal Society*, 1, 267, Edinburgh, 1788

page 17 'each stratum...' John Phillips, *Memoirs of William Smith*, 1844

page 23 'geologic time...' C. D. Walcott, 'Geologic Time, as indicated by the sedimentary rocks of North America', *Journal of Geology*, Vol.1, 1893

page 24 'I came into the room...' A. S. Eve, *Rutherford*, Macmillan, New York, 1939

### Chapter 2

page 34 'it is just as if...' Alfred Wegener, *The Origins of Continents and Oceans*, translated from the 4th revised German ed. of 1929 by J. Biram, with an introduction by B. C. King, Methuen, London, 1966

page 53 'the structure of the sea floor...' Allan J. Cox (ed.), *Plate Tectonics and Geomagnetic Reversals*, Freeman & Co, San Francisco, 1973

# Glossary
. . . . . . . . . . . . . . .

**andesite** A fine-grained volcanic rock, named after the volcanoes of the Andes, mainly composed of feldspar, pyroxene, and sometimes quartz and olivine – its composition is defined as intermediate.

**asthenosphere** The weak part of the mantle which immediately underlies the lithosphere.

**atmosphere** Layer of gases which encloses the Earth – almost all the mass of the atmosphere is in the bottom 30 kilometres.

**basalt** A fine-grained volcanic rock mainly composed of feldspar, pyroxene and sometimes olivine – its composition is defined as basic.

**Benioff zone** An inclined zone of earthquakes in subduction zones.

**black smoker** A column of hot muddy water which gushes out of the sea floor.

**carbonaceous chondrite** A type of meteorite which has a very similar composition to the solar nebula, and is rich in carbon and water.

**Cenozoic** 'Time of young life' which spans the period between 65 and 2 million years ago.

**chondrite** Common meteorite containing spherical structures called chondrules.

**continent** Portion of the Earth's surface which is underlain by continental crust and usually stands above sea level.

**continental crust** Outer layer of the solid Earth beneath the continents, which is on average about 35 kilometres thick (but thicker beneath mountainous regions) and is rich in the minerals quartz and feldspar.

**continental shelf** Margins of the oceans, extending up to a few hundred kilometres offshore, where the water depth is less than 200 metres.

**convection** Pattern of flow in a cooling fluid in which hot parts rise and cold parts sink.

**core** The central part of the Earth where the Earth's magnetic field is generated, composed mainly of iron and nickel; the inner core is solid, but the outer core is molten.

**crust** Outermost layer of the solid Earth with a different composition to the underlying layers

**drift** Chaotic deposits of rock fragments and other detritus, transported and eventually left behind by glaciers.

**echo-sounder** Device for measuring the depth of the oceans.

**elastic rebound** Property of springiness of the Earth's crust, so that during an earthquake crust 'snaps' back like a stretched rubber band.

**eukaryote** A single-celled organism which contains a nucleus and internal cell organelles (compartments) and exploits oxygen.

**Euler pole** Axis of rotation which describes the relative motion of two plates.

**fault** Break in the Earth's crust – sudden movement on a fault triggers an earthquake.

**feldspar** A silicate mineral which is common in igneous rocks, particularly granites.

**fluid** A substance which is capable of flowing – solid materials can be fluids.

**garnet** A silicate mineral which is commonly found in metamorphic rocks.

**glacial period** An interval during an Ice Age when temperatures are generally coldest and ice sheets are extensive

**glacial maximum** The coldest time in a glacial period when ice sheets are at their maximum extent.

**Gondwanaland** A super-continent which existed about 200 million years ago when the continents of Africa, South America, Australia, India and Antarctica were joined together.

**granite** A coarse-grained igneous rock, commonly found in the continental crust and composed mainly of the minerals quartz, feldspar and mica.

**greenhouse gas** A gas whose presence has the effect of making the atmosphere warmer than it would otherwise be without the gas – carbon dioxide and water vapour are important examples.

**half-life** Time taken for a population of radioactive elements to decay to half their original number.

**hot spot** Site of volcanism away from the edges of plates.

**Ice Age** Period in Earth history when large ice sheets exist on land – we are in an Ice Age today.

**igneous rock** Rock that has cooled from a molten state.

**interglacial period** An interval during an Ice Age when temperatures are warmest and ice sheets are at their minimum extent.

**isotope** One of two or more forms of an element differing in atomic weight.

**lava** Molten rock (or solidified remains) which flows out from a volcano during a volcanic eruption.

**lithosphere** Strong outer part of the solid Earth which forms a plate – consists of both crust and part of the underlying mantle, and is usually about 100 kilometres thick.

**mantle** Portion of the Earth between the crust and core – the top of the mantle is mainly composed of the minerals olivine and pyroxene.

**mantle plume** A hot upwelling region in the mantle which rises to near the Earth's surface, usually triggering volcanism.

**Mesozoic** 'Time of middle life' – period in Earth history between 65 and 250 million years ago.

**metamorphic rock** A rock which has undergone changes due to the effects of temperature and pressure after the rock first formed.

**meteorite** Body of rock, which after drifting in space, falls to Earth.

**mica** A silicate mineral which has a platy or sheet-like nature – sheets of white mica are often used as fire doors in stoves.

**mid-ocean ridge** A linear zone of shallowing in the middle of oceans – new ocean crust is created by volcanic eruptions along the crest of the mid-ocean ridge.

**nebula(e)** Region(s) of gas and dispersed solid particles in space.

**oceanic crust** Crust beneath the oceans which is about 7 kilometres thick, consisting of three principal layers: a top covering thin layer of sedimentary rocks; a middle layer of fine-grained volcanic rock (basalt); and a bottom layer of coarse-grained igneous rock.

**ocean trench** Long deep depression in the ocean floor where a plate bends down and sinks back into the Earth's interior.

**olivine** A silicate mineral which is rich in magnesium and iron and is commonly found in the upper parts of the mantle.

**ophiolite** A geological formation, found on land, which comprises the three layers of the ocean crust, resting on mantle rocks.

**P wave** A form of seismic vibration, generated by an earthquake, which travels through the Earth's interior – the motion of the vibrating rock is a bit like the vibration of a spring.

**Palaeozoic** 'Time of old life' – the period in Earth's history between 250 and c. 550 million years ago.

**Pangea** A supercontinent which straddled the Earth about 250 million years ago, when all the continents were joined together.

**plates** The curved, rigid parts of Earth's outer shell (synonymous with lithosphere) which move relative to each other.

**plate tectonics** The behaviour of plates.

**post-glacial rebound** The process by which the surface of the Earth returns to its original shape after being depressed by the weight of vast ice sheets.

**Precambrian** Portion of Earth history which predates the Palaeozoic.

**prokaryote** A single-celled organism which has no nucleus or internal organelles (compartments) and often does not rely on oxygen for survival.

**pyroxene** A silicate mineral which is commonly found in igneous rocks and the upper part of the Earth's mantle.

**quartz** A silicate mineral composed entirely of silicon and oxygen.

**Rodinia** A supercontinent which straddled the Earth about 750 million years ago, when all the continents were joined together.

**S wave** A form of seismic vibration, generated by an earthquake, which travels through the Earth's interior –

the motion of the vibrating rock is a bit like the sideways motion of a snake.

**schist** A medium-grained metamorphic rock, usually containing platy minerals such as mica, which tends to break along numerous parallel planes

**sedimentary rock** Rock made up of older rock fragments which were transported at the surface of the Earth, usually in water or by the wind, and then deposited in layers.

**silicate** A type of mineral, which makes up most of the Earth's crust and mantle, composed largely of silicon and oxygen.

**solar nebula** Nebula out of which the Solar System was created.

**strata** The layers of sedimentary rock.

**stromatolite** Rock formed by the action of algae, composed of numerous thin hummocky layers of carbonate minerals and rock detritus.

**subduction (zone)** Process (or region of the Earth) by which the lithospheric plate sinks back into the Earth's mantle.

**transform fault** A fault between two plates where the plates slide horizontally past each other.

**tomography (seismic)** Producing images of the Earth's interior using earthquake vibrations.

**unconformity** The contact between two rock formations which is the result of substantial erosion before the deposition of the younger 'sedimentary' formation.

**volcanic arc** Chain of volcanoes which lies above a subduction zone.

**volcanic rock** Rock produced as a result of volcanic activity – usually formed when molten lava cools.

# ACKNOWLEDGEMENTS

The television series would not have been possible without the talented team of producers, Robin Brightwell, Cynthia Page, Danielle Peck, Isabelle Rosin and Simon Singh, and our creative and expert researchers, Liz Drake, Chris Nicholas and Duncan Copp. We have all benefited from the help and advice of our Executive Producer Richard Reisz, Unit Manager Phil Checkland, and Production Manager Sue Crane. Many other people made important creative contributions to the television series, especially our film editors, Jon Bignold, Christy Hanna, Helen Walker and Alice Forward, and our Designer Andrew Sides, as well as Alex Hope and the team from the Moving Picture Company. We would also like to thank Evonne Francis, Alex Branson, Cathy Walker and Paul Ralph for their invaluable help in the office and on location. A project of this scale needs powerful friends, and in this regard we all owe a debt of gratitude to Jana Bennett, former Head of BBC Science, who has always been an enthusiastic supporter of *Earth Story*, and without whose backing it would never have seen the light of day.

# PICTURE CREDITS

# Index